181
Advances in Polymer Science

Advances in Polymer Science

Recently Published and Forthcoming Volumes

Interphases and Mesophases in Polymer Crystallization III
Volume Editor: Allegra, G.
Vol. 191, 2005

Microseparation of Block Copolymers II
Volume Editor: Abetz V.
Vol. 190, 2005

Microseparation of Block Copolymers I
Volume Editor: Abetz V.
Vol. 189, 2005

Intrinsic Molecular Mobility and Toughness of Polymers II
Volume Editor: Kausch H. H.
Vol. 188, 2005

Intrinsic Molecular Mobility and Toughness of Polymers I
Volume Editor: Kausch H. H.
Vol. 187, 2005

Polysaccharides I
Structure, Characterisation and Use
Volume Editor: Heinze T. T.
Vol. 186, 2005

Advanced Computer Simulation Approaches for Soft Matter Science II
Volume Editors: Holm C., Kremer K.
Vol. 185, 2005

Crosslinking in Materials Science
Vol. 184, 2005

Phase Behavior of Polymer Blends
Volume Editor: Freed K.
Vol. 183, 2005

Polymer Analysis · Polymer Theory
Vol. 182, 2005

Interphases and Mesophases in Polymer Crystallization II
Volume Editor: Allegra G.
Vol. 181, 2005

Interphases and Mesophases in Polymer Crystallization I
Volume Editor: Allegra G.
Vol. 180, 2005

Inorganic Polymeric Nanocomposites and Membranes
Vol. 179, 2005

Polymeric and Inorganic Fibres
Vol. 178, 2005

Poly(arylene ethynylene)s
From Synthesis to Application
Volume Editor: Weder C.
Vol. 177, 2005

Metathesis Polymerization
Volume Editor: Buchmeiser M. R.
Vol. 176, 2005

Polymer Particles
Volume Editor: Okubo M.
Vol. 175, 2005

Neutron Spin Echo in Polymer Systems
Authors: Richter D.,
Monkenbusch M., Arbe A., Colmenero J.
Vol. 174, 2005

Advanced Computer Simulation Approaches for Soft Matter Sciences I
Volume Editors: Holm C., Kremer K.
Vol. 173, 2005

Microlithography · Molecular Imprinting
Vol. 172, 2005

Polymer Synthesis
Vol. 171, 2004

NMR · 3D Analysis · Photopolymerization
Vol. 170, 2004

Long-Term Properties of Polyolefins
Volume Editor: Albertsson A.-C.

Interphases and Mesophases in Polymer Crystallization II

Volume Editor: Giuseppe Allegra

With contributions by
A. Abe · F. Auriemma · S. Bracco · A. Comotti · P. Corradini
W. H. de Jeu · C. De Rosa · H. Furuya · T. Hiejima · Y. Kobayashi
L. Li · R. Simonutti · P. Sozzani · Z. Zhou

Springer

This series presents critical reviews of the present and future trends in polymer and biopolymer science including chemistry, physical chemistry, physics and material science. It is adressed to all scientists at universities and in industry who wish to keep abreast of advances in the topics covered.

As a rule, contributions are specially commissioned. The editors and publishers will, however, always be pleased to receive suggestions and supplementary information. Papers are accepted for "Advances in Polymer Science" in English.

In references Advances in Polymer Science is abbreviated *Adv Polym Sci* and is cited as a journal.

The electronic content of *Adv Polym Sci* may be found at springerlink.com

ISBN 978-3-642-06465-4 e-ISBN 978-3-540-31576-6

ISSN 0065-3195

DOI 10.1007/b107168

Springer is a part of Springer Science+Business Media
springeronline.com

Softcover reprint of the hardcover 1st edition 2005

Cover design: *Design & Production* GmbH, Heidelberg

Printed on acid-free paper 02/3141 YL – 5 4 3 2 1 0

Volume Editor

Prof. Giuseppe Allegra
Dipartimento di Chimica
Materiali e Ingegneria Chimica "Giulio Natta"
Politecnico di Milano
Via L. Mancinelli, 7
20131 Milano

Editorial Board

Advances in Polymer Science Also Available Electronically

For all customers who have a standing order to Advances in Polymer Science, we offer the electronic version via SpringerLink free of charge. Please contact your librarian who can receive a password or free access to the full articles by registering at:

springerlink.com

If you do not have a subscription, you can still view the tables of contents of the volumes and the abstract of each article by going to the SpringerLink Homepage, clicking on "Browse by Online Libraries", then "Chemical Sciences", and finally choose Advances in Polymer Science.

You will find information about the

- Editorial Board
- Aims and Scope
- Instructions for Authors
- Sample Contribution

at springeronline.com using the search function.

Preface

Polymer crystallisation is a field of science whose widespread practical and technological implications add to its scientific relevance. Unlike most molecular substances, synthetic polymers consist of long, linear chains usually covering a broad distribution of molecular lengths. It is no surprise that only rarely may they give rise to regularly shaped crystals, if at all. As a rule, especially from the bulk state, polymers solidify as very tiny crystals interspersed in an amorphous matrix and randomly interconnected by disordered chains. How do these crystals form? Do they correspond to a state of thermodynamic equilibrium, or are the chains so inextricably entangled that equilibrium is virtually impossible to reach? There is currently a widespread consensus on the latter conclusion, which only makes the problem more interesting as well as more difficult to handle. The perspective at the base of the present endeavour can be summarised with two questions: What are the key structural steps from the original non-crystalline states to the semi-crystalline organization of the polymer? Do these different stages influence the resulting structure and to what degree?

As demonstrated by the collection of review articles published within three volumes of *Advances in Polymer Science* (Volumes 180, 181 and 191), this problem may be approached from very different sides, just as with the related topic of polymer melting, for that matter. Morphological and atomistic investigations are carried out through the several microscopic and scattering techniques currently available. X-ray, neutron and electron diffraction also provide information to unravel the structure puzzle down to the atomistic level. The same techniques also allow us to explore *kinetic* aspects. The fast development of molecular simulation approaches in the last few decades has given important answers to the many open problems relating to kinetics as well as morphology; in turn, statistical-mechanical studies try to make sense of the many experimental results and related simulations. In spite of several successes over 60 years or more, these studies are still far from providing a complete, unambiguous picture of the problems involved in polymer crystallisation. As one of the authors (an outstanding scientist as well as a very good friend) told me a couple of years ago when we started thinking about this project, we should not regard this book as the solution to our big problem – which it is not – but rather as a sort of "time capsule" left to cleverer and better-equipped scientists of generations to come, who will make polymer crystallisation completely clear.

Thanks to all the authors for making this book possible. Here I cannot help mentioning one of them in particular, Valdo Meille, who helped with planning, suggesting solutions and organising these volumes. Thank you, Valdo, your intelligent cooperation has been outstandingly useful.

Milan, February 2005 Giuseppe Allegra

Contents

Contents of Volume 180

Interphases and Mesophases in Polymer Crystallization I

Volume Editor: Giuseppe Allegra

ISBN: 3-540-25345-9

Contents of Volume 191

Interphases and Mesophases in Polymer Crystallization III

Volume Editor: Giuseppe Allegra
ISBN: 3-540-28280-7

Adv Polym Sci (2005) 181: 1–74
DOI 10.1007/b107169

Published online: 30 June 2005

Solid Mesophases in Semicrystalline Polymers: Structural Analysis by Diffraction Techniques

Finizia Auriemma (✉) · Claudio De Rosa · Paolo Corradini

Dipartimento di Chimica, Università di Napoli "Federico II", via Cintia, 80126 Napoli, Italy
Finizia.Auriemma@unina.it, Claudio.Derosa@unina.it, Paolo.Corradini@unina.it

Abstract Crystalline polymers may be affected by various kinds and amounts of structural disorder. Lack of order in polymer crystals may arise from the presence of defects in the chemical constitution, configuration and conformation and defects in the mode of packing of chains inside the crystals. Polymeric materials characterized by long range order in the parallel arrangement of chain axes and a large amount of structural disorder may be considered as solid mesophases. The different kinds and amount of disorder present in the solid mesophases of semicrystalline polymers and the possible types of solid mesophases are discussed in terms of idealized limit models of disorder. These models imply maintenance of long-range positional order, at least along one dimension, of structural features which are not necessarily point centered. Structural features, which are not point centered, are for instance chain axes, the center of mass of special groups of atoms or bundles of chains and so on. Structural aspects emerging from the X-ray diffraction analysis of several solid mesophases discussed so far in the literature are reviewed, in the light of the present analysis.

Keywords Solid mesophases · Disordered structures · Diffuse scattering · X-ray diffraction · Structure modeling

1 Introduction

The term “mesomorphic” was proposed by Friedel in 1922 for materials of “middle” (Greek: mesos) “form” (Greek: morphe) to address materials in a condensed phase having intermediate characteristics between liquids and crystals [1]. More generally, the term “mesomorphic” may be used to address all the states of matter, which may be considered as intermediate between the crystalline and the liquid (or amorphous) state, as for instance “positionally disordered” crystals or “orientationally ordered” liquids and/or glasses [2].

A tentative classification of mesophases was reported by Wunderlich [2], who divided mesophases into six different types of phases: liquid crystals (LC), condis crystals (CD), plastic crystals (PC) and the corresponding LC, CD and PC glasses.

Liquid crystals are materials characterized by long-range orientational order of molecules as in crystals but absence of three-dimensional positional order as in liquids. In these “positionally disordered” crystals or “orientationally ordered” liquids, large-scale molecular motion is possible [3–5]. The name “liquid crystals” was given by Lehman in 1907 [6] because of their optical anisotropy and the liquid-like flow of these materials.

Plastic crystals are characterized by orientational disorder but positional order of the structural motif. Molecules of plastic crystals are close to spherical, which are generally packed in body- or face-centered cubic structures [7]. Typical examples are provided by the structure of ball-like hydrocarbon molecules as adamantane and norbornane. The name “plastic crystals” derives from the softness and easy of deformation of these materials, due to the large number of slip planes in close packed structures [5, 7].

The term "condis crystals" has been used to identify "conformationally disordered" crystals, i.e. structures characterized by disorder in the conformation of molecules [2].

Plastic crystals and condis crystals are two kinds of solid mesophases, whereas liquid crystals are essentially liquids. The differences between these three mesophases are largely based on the geometry of the molecules: the molecules of liquid crystals always have a rigid, mesogenic group which is rod- or disk-like and causes a high activation barrier to rotational reorientation [3, 5]. The molecules of plastic crystals are compact and rather globular, so that there is no high activation barrier to their reorientation [5, 7]. Condis crystals consist of flexible molecules which can easily undergo changes in the conformation without losing positional or orientational order [2].

According to Wunderlich [2], positionally disordered or LC-glasses, orientationally disordered or PC-glasses and conformationally disordered or CD-glasses, are terms identifying glasses obtained by quenching liquid crystals or the melt of plastic crystals and condis crystals, respectively, at temperatures below the glass transition, preventing crystallization.

As discussed in [2], in a condis crystal cooperative motion between various conformational isomers is permitted, whereas in the CD-glass this motion is frozen, but the conformationally disordered structure remains. In the case of polymers, a condition for formation of condis crystals is that the macromolecules exist in conformational isomers of low energy, which leave the macromolecules largely in extended conformations so that the parallelism of chain axes is maintained.

Condis crystals include a large number of solid mesophases of polymers, as for instance the high-temperature crystalline forms of 1,4-*trans*-poly(1,3-butadiene) [8, 9] and poly(tetrafluoroethylene) [10–12]. In both cases, the chains are conformationally disordered; long-range order is maintained as far as the parallelism of chain axes and the pseudo-hexagonal placement of chain axes are concerned.

More in general, solid mesophases not only include crystalline forms of polymers containing a large amount of disorder in the conformation of chains and long-range order in the position of chain axes as in condis crystals, but also crystalline polymers characterized by disorder in the lateral packing of conformationally ordered chains [13, 14].

A notable difference between the solid mesomorphic forms and the ordered liquid mesophases is that the solid mesomorphic forms are crystalline modifications generally characterized by the typical feature of crystalline order, that is, the packing of parallel chains. The long-range order in the correlations between the atoms of the parallel chains is absent because of the presence of disorder. The long-range order may be lost in one or two dimensions, for example, when conformationally disordered chains are packed with long-range order in the position of chain axes or for conformationally ordered chains packed with a high degree of disorder in the lateral packing [13, 14]. The presence of these kinds of disorder generally prevents the definition of a unit cell. Typical features in the X-ray diffraction patterns of solid mesophases are the presence of a large amount of diffuse scattering and a few (if any) Bragg reflections [13].

Solid mesophases are extremely frequent in polymers. For instance, at high pressure the orthorhombic form of polyethylene transforms into a hexagonal mesophase, characterized by a high degree of disorder (conformational disorder) [15–17]. In some cases, the amorphous phase may transform into a mesophase by stretching at temperatures lower than the glass transition temperature (e.g. in poly(ethylene terephthalate) [18–20], syndiotactic polystyrene [21, 22], nylon 6 [23, 24]) or by quenching the melt at low temperatures (isotactic [25–27] and syndiotactic [28, 29] polypropylene). Copolymers of ethylene/propylene with a propylene content in the range of 15–35 mol % are amorphous at room temperature and in the unstretched state, but crystallize into a pseudo-hexagonal mesomorphic form, by cooling at low temperatures or by stretching at room temperature [30–38]. In the case of atactic polyacrylonitrile, the crystalline pseudo-hexagonal polymorph is actually a mesophase [39–48]. The already mentioned high-temperature forms of poly(tetrafluoroethylene) [10–12, 49–54] and of 1,4-*trans*-poly(1,3-butadiene) [8, 9, 55–58] may also be considered mesophases. Some, but not all, of the above listed mesophases are "condis crystals" in the sense of Wunderlich [2].

It is very difficult to classify the various kinds of solid mesophases of polymers described so far in the literature in a simple and general way, because in the crystals different kinds and different degrees of disorder may be present at the same time; often it is not easy to identify which kind of disorder mainly characterizes a given mesophase.

Because of the presence of structural disorder, the X-ray diffraction patterns of mesophases show a large amount of diffuse scattering and need a special care for a quantitative evaluation. Paracrystalline distortions of the lattice [59, 60] usually affect the shape and the width of the diffraction peaks to a large extent. The analysis of disorder necessarily implies a multidisciplinary approach, in order to unravel the complicated nature of disorder in disordered crystalline materials [61].

In this review, we attempt to present the subject of solid mesophases of polymers mainly in relation to structural aspects emerging from diffraction experiments. The relationships between the structure and properties of these materials are analyzed and related to the amount and kind of disorder present in the crystals. The study of disorder in solid mesophases, indeed, allows a deep comprehension of phenomena subtending their chemical and physical properties.

In line with arguments of [13], with the terms "solid mesophase" we identify states of matter falling "in between" amorphous and crystalline states, characterized by long-range order in the parallel arrangement of chain axes.

We will attempt to outline possible classes of solid mesophases, not based on the different kinds and amount of disorder present in the crystals, but rather in terms of idealized limit models of disorder. These models imply maintenance of long-range positional order, at least along one dimension, of structural motifs, which are not necessarily point centered [13].

2 Disorder in Crystals of Low Molecular Mass Molecules: Limit Models of Mesophases

A model of ideal crystal consists in a three-dimensional array of identical units. Such an "ideal" crystal gives rise to calculated diffraction patterns consisting of only discrete diffraction peaks (called Bragg reflections). Thermal movements are disregarded in this definition.

The definition of crystal as a periodic array of identical motifs corresponds to the mathematical concept of lattice. Under this assumption, crystallography has developed powerful methods for allowing crystal structure determination of different materials, from simple metals (where the asymmetric unit may be the single atom) to protein crystals (containing thousands of atoms per cell) [62].

However, real crystalline materials only approximate this ideal model. Departures from the ideal periodic array of identical units may usually occur in a variety of different ways and implies presence of disorder [63, 64]. The presence of disorder in the crystals gives rise to a diffuse, continuous scattering in diffraction patterns. Disorder, indeed, may cause broadening of Bragg reflections and a decrease of the maximum height of the peak intensity; for a given Bragg reflection, however, the intensity which is lost because of the presence of disorder is found in the background and represents the diffuse scattering [61, 63, 64].

First of all, disorder in crystals arises from thermal motion of atoms around their equilibrium positions (thermal disorder). Disorder may also originate from the presence of defects in the lattice, due, for instance, to substitution of atoms (or group of atoms) with atoms (or group of atoms) of a different chemical nature, inclusion of structural motifs in interstitial lattice positions, or when structural units may pack in the same basic crystal lattice in different orientations or in a different conformation [63, 64]. The presence of defects in the crystals, in turn, may induce local deformation of the lattice (due to small displacements of atoms from their average positions), which helps relaxation of local stresses, and causes diffuse scattering.

According to Hosemann [59, 60], the lattice distortions, in addition to those due to thermal vibrations of atoms, may be classified as of a) first kind if the long-range periodicity is preserved with respect to the average positions over all the lattice points, and of b) second kind if the position of each lattice

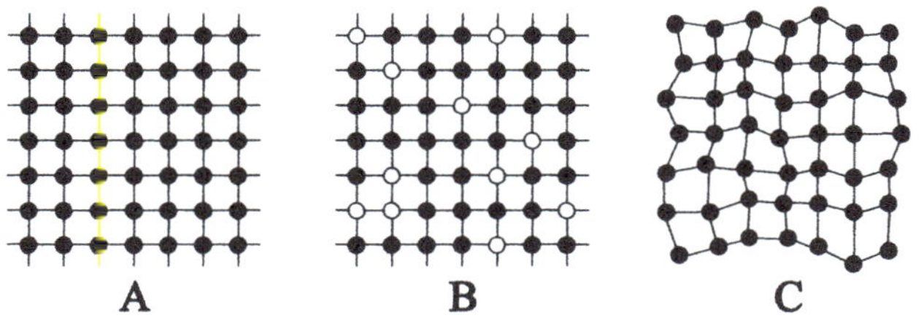

Fig. 1 (**A**) an ideal two-dimensional lattice. (**B**) first-kind distortions implying substitution type disorder; filled and empty balls indicate different chemical structural units. (**C**) Lattice distortions of second kind (paracrystal)

point deviates statistically only with respect to its nearest neighbors instead of respect to the ideal lattice points. As an example, lattice distortions of the first and second kind in a bidimensional lattice are schematically represented in Fig. 1.

The presence of lattice distortions of the first kind induces a decrease of the intensities of X-ray diffraction reflections with increasing diffraction order, whereas the width of reflection spots does not vary. Lattice distortions of the second kind, instead, result in both a diminution of reflection intensity and an increase in reflection breadth, with increasing reflection order [59, 60]. They are also called paracrystalline distortions [60].

3 Disorder in Semicrystalline Polymers: Limit Models of Mesophases

3.1 General Principles

Disregarding thermal motion, the concept of an ideal crystal requires long-range positional order for all atoms. For polymers, this ideal condition implies an infinite chain length, regular constitution, regular configuration and regular conformation [65].

The requirement of a regular constitution implies that all repeating units have the same chemical structure. The requirement of a regular configuration implies that whenever a monomeric unit may assume different configurations, the succession of configurations must be regular. The requirement of a regular conformation implies that the conformation assumed by polymer chains of a given constitution and configuration in the crystalline state can be defined as a succession of structural units which occupy geometrically (not necessarily crystallographically) equivalent positions with respect to a chain axis (equivalence principle) [65, 66]. The chain axis is generally parallel to a crystallographic axis of the crystal.

The geometrical equivalence of structural units along an axis allows defining types of geometrical symmetry that a linear macromolecule may achieve in the crystalline sate. The conformation of a macromolecule is generally defined in terms of its symmetry and, precisely, of the line repetition symmetry group [65, 67–69].

In the framework of the allowed symmetries, that is under the restrictions imposed by the need of having an axis of repetition, the conformation of the chain will tend to approach one of minimum conformational energy for the isolated chain [65, 70–72]. Packing effects generally do not influence the conformation of the chains as long as the conformational energy of the isolated chain corresponds to a deep energy minimum. Crystal packing effects may influence the choice among conformations of nearly equal energy for the isolated chain [73].

Of course, none of the above requirements concerning infinite chain length, the regularity in constitution, configuration and conformation, can be fulfilled in the crystals of real polymers, which are then disordered [14, 65].

Some kinds of disorder are as follows: First of all, polymers (as well as oligomers, in general) are materials consisting of molecules of non-uniform molecular mass [74]. At variance with biological macromolecules, as nucleic acids and polypeptides, which are known to be uniform at least with respect to the molecular structure of the core (that is, sequence arrangement of constitutional units, stereoregularity and overall conformation), the macromolecular chains in a given polymer may be similar, but, certainly, not all alike [74, 75].

In polymers, examples of constitutional faults with respect to an idealized model include mistakes in the head-to-tail succession of monomeric units, the presence of units derived from a different monomer and in any case, the chain ends [65].

Copolymers of different monomeric units that are able to cocrystallize in the same lattice may be considered as a possible exception to the need of regularity in the chemical constitution. For instance, in the case of vinyl polymers it is possible to accommodate into the crystalline lattice different comonomeric units having lateral groups with different shape and dimensions. Copolymers of butene and 3-methylbutene [76] or copolymers of styrene and *o*-fluorostyrene [77] are crystalline over the whole range of composition, provided that the copolymers are isotactic whereas propene and 1-butene also are able to cocrystallize at any composition in syndiotactic copolymers [78, 79]. Isomorphism of comonomeric units also occurs in copolymers of acetaldehyde and *n*-butyraldeide [80], which are crystalline over the whole range of composition. Analogous isomorphism of comonomeric units has been observed in *trans*-1,4-copolymers of butadiene and 1,3-pentadiene [81] and in copolymers of vinylidenfluoride and vinylfluoride [82].

An example of a configurational mistake is the possible presence, in an isotactic polymer, of racemic diads in place of all *meso*-diads. In several cases, the presence of configurational defects in chains of stereoregular polymers does not prevent crystallization; their possible inclusion in the crystals may be energetically feasible, provided that the portion of chain close to the defect adopt a low-energy conformation which does not disrupt the packing of close neighboring chains [83–86]. Moreover, even stereoirregular polymers are able to crystallize; classical examples are poly(vinyl alcohol) [87, 88], poly(vinyl chloride) [89–94] and poly(acrylonitrile) [39–48, 95–97].

For polymers with a regular constitution and configuration, the conformation adopted by chains in the crystalline state is generally regular. However, some three-dimensional, long-range crystalline order may be maintained even when disorder is present in the conformation of polymer chains. The term conformational isomorphism refers to the more or less random occurrence in the same lattice site of different, but almost isoenergetic, conformers of the same portion of a molecule [58]. A classification of the possible cases of conformational isomorphism in polymers can be made on the basis of the following two possibilities [98, 99]:

1. The chain may assume, in the same crystal and more or less at random, different (although nearly isosteric) conformations.
2. If the geometry of the main chain is fixed, the lateral groups may assume different conformations in the same crystal, more or less at random.

Crystalline polymers characterized by disordered conformations of the chains are, for instance, polytetrafluoroethylene [10–12, 49–54], *cis*-1,4-poly (isoprene) [100–102] and *trans*-1,4-poly(1,3-butadiene) [8, 9, 55–58]. In these cases, disorder does not destroy the crystallinity because of the similar shape of the various conformational units. The occurrence of cases of conformational isomorphism of the first kind demonstrates that a polymer chain can remain straight, as if it was constrained to run inside the walls of a tight cylinder, while its conformational freedom remains of the same order of magnitude as that in the melt.

The case of conformational isomorphism of lateral groups occurs, for instance, in the crystal structure of isotactic poly(S-3-methyl-pentene), characterized by the presence of fourfold helices of only one sense (left-handed), with lateral groups which may adopt statistically two conformations of minimum energy [99, 103].

In addition, conformational disorder in polymer crystals may give rise to point and line defects which are tolerated in the crystal lattice at a low cost of free energy as kinks [104, 105], jogs [106, 107] and dislocations [108, 109]. Such crystallographic defects arise whenever portions of chain adopt conformations different from the conformation assumed by the chains in the crystal state [99], and have been widely discussed in the literature, in the case of polyethylene [108, 109] and some aliphatic polyamides [99, 106]. Point and line defects may be important in determining some physical properties of polymeric materials [110, 111].

Lack of order may arise not only from the presence of defects in the chemical constitution, configuration and conformation but also by defects in the mode of packing of chains inside the crystals. Even in the case of regular constitution, configuration and conformation, disorder may be present in the crystals due to the presence of defects in the mode of packing [13, 14, 65]. Generally, disorder in the packing of polymer chains may occur while a long-range positional order of some structural feature is maintained [13].

3.2 Classification of Disordered Systems

Different kinds of disorder may affect differently the X-ray diffraction pattern of crystalline polymeric materials. Depending on the features present in the X-ray fiber diffraction patterns, it is useful to distinguish here the following main classes of disordered systems [13]:

i) The long-range, three-dimensional periodicity is maintained only for some characterizing points of the structure. In these cases, there is only a partial three-dimensional order, and the X-ray diffraction patterns present sharp reflections and diffuse halos. Disordered structures belonging to this class may be characterized, for instance, by disorder in the positioning of right- and left-handed helical chains, or disorder in the up-down positioning of conformationally ordered chains or disorder in the stacking of ordered layers of chains along one crystallographic direction [13, 14, 65].
ii) The long-range positional three-dimensional order is maintained only for some structural features which are not point centered, for example, the chain axes. Structures characterized by conformationally disordered chains with a long-range order in the position of the chain axes belong to this class. The X-ray fiber diffraction pattern of this kind of disordered structure is characterized by sharp reflections on the equator and diffuse halos on the layer lines.
iii) The long-range positional order of some features is maintained only in two or in one dimension, for example along the chain axis. Examples may be provided by structures characterized by conformationally ordered parallel chains with disorder in the lateral packing of the chains. Short-range correlations between atoms of neighboring chains may still be present, but long-range order is maintained only along the axis of each chain. The X-ray fiber diffraction pattern is characterized by well-defined layer lines presenting diffuse scattering.

In all cases, the X-ray diffraction patterns present Bragg peaks and diffuse scattering corresponding to precise zones of the reciprocal lattice. The analysis of Bragg peaks provides information about the average crystal structure as for instance average atomic positions, mean-squared atomic displacements, site occupancy. Analysis of diffuse scattering may allow accessing information about how structural motifs (atoms or groups of atoms) interact [63, 112]. Position, shape and intensity of the diffuse scattering, indeed, depend on the kind and amount of disorder present in the structure.

According to [13], a large part of the crystal structures that fall in case i) may be conveniently described adopting the concepts of "limit-ordered" and "limit-disordered" model structures. A limit-ordered structure defines an ideal model with a perfect three-dimensional, long-range order (periodicity) of all atoms in the structure. A limit-disordered model structure describes an "ideal" model characterized by the maintenance of three-dimensional long-range order (periodicity) of only some points of the structure, and a full disorder in the positioning of other points of the structure. In real cases, the true structures are, in general, intermediate between fully ordered and fully disordered limit models.

As an example, the structure of the α form of isotactic polypropylene may be described with reference to a limit-ordered model (defined α_2-form) [113] and a limit-disordered model (defined α_1-form) [114], shown in Fig. 2B and C, respectively.

Disorder in Fig. 2C arises from the fact that in the crystals 3/1 helices of isotactic polypropylene of a given chirality (left- or right-handed, see Fig. 2A) may be oriented in two opposite directions, i.e. with C – C bonds connecting the methyl groups either pointing toward the positive direction ("up") of an oriented axis (z) parallel to chain axis, or in the negative direction ("down") of z axis. In the limit-ordered structural model of Fig. 2B "up" and "down" helices follow each other according to a well-defined pattern (primitive unit cell, space group $P2_1/c$) [113]; in the limit-disordered model structure of Fig. 2C up and down chains having the same chirality may substitute each other in the same site of the lattice and the resulting model may be described, on the average, in terms of a C-pseudo-centered unit cell (space group $C2/c$) [114]. The real crystalline modifications of isotactic polypropylene are intermediate between the two limit-ordered and limit-disordered models, the degree of up/down disorder being dependent on the thermal and mechanical history of the sample [113–120]. Up/down disorder of isotactic polypropylene in the α polymorph corresponds to a case of isomorphism where non-equivalent isosteric orientations of the same conformation may substitute vicariously for each other in the lattice and is quite common in the crystalline polymorphs of isotactic vinyl polymers [58].

In the cases ii) and iii), degrees of disorder higher than in case i) are present. Few Bragg reflections are present in the diffraction patterns and dif-

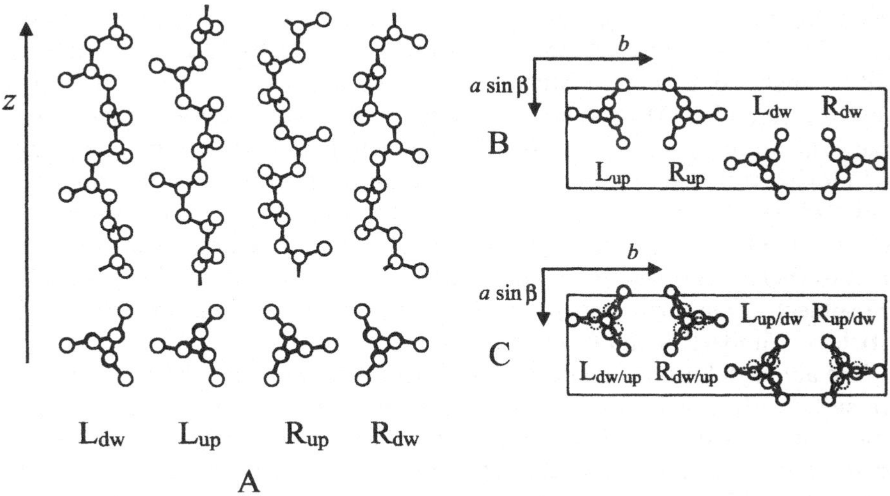

Fig. 2 (**A**) Chain conformation of isotactic polypropylene in the crystalline state. Symbols "R" and "L" identify right- and left-handed helices, respectively, in 3/1 conformations. Subscripts "up" and "dw" ("dw" standing for "down") identify chains with opposite orientation of C – C bonds connecting tertiary carbon atoms to the methyl groups along the *z*-axis (**B**) Limit-ordered model structure ($\alpha 2$ modification, space group $P2_1/c$) [113] (**C**) Limit-disordered model structure ($\alpha 1$ modification, space group $C2/c$) [114]. In the $\alpha 2$ modification "up" and "down" chains follow each other according to a well-defined pattern. The $\alpha 1$ modification presents a complete disorder corresponding to a statistical substitution of "up" and "down" isomorphic helices

fuse scattering may cover even large zones of the reciprocal space. It is not generally possible to define a unit cell, but only average periodicities along some directions (for instance parallel to chain axes) can be established. These solid phases may be considered as intermediate between perfectly crystalline and completely amorphous phases and can be called "mesomorphic forms" or "solid mesophases" [13].

This allows identifying two large classes of solid mesophases: class ii) where long-range periodicity along three-dimensions may be still defined, in spite of the presence of large amounts of disorder and class iii) where periodicity may be defined only in one or two dimensions.

An important feature that characterizes crystalline materials of class i) is a low amount of disorder implying large changes of enthalpy and entropy of melting. Solid mesophases of class ii) and iii) are instead largely disordered. The melting entropy is generally low in these systems and very often chains experience only weak interactions with the surrounding lattice environment.

Polymeric materials belonging to class i) will not be treated in this review, for which we refer to extensive treatments as those of [13, 14, 65]. This review is essentially focused on crystalline materials with a large amount of disorder, belonging to classes ii) and iii). The basic structural organization in solid mesophases belonging to these two classes, as revealed by diffraction studies, will be analyzed. Structural studies performed using solid-state NMR spectroscopy, IR and RAMAN spectroscopy, inelastic neutron scattering, thermal analysis, conformational analysis, X-ray modeling and other complementary techniques, will also be discussed in addition to the structural information provided by diffraction techniques. Thermodynamic aspects concerning the relative stability of these mesophases will also be pointed out. The issue related to thermodynamics of solid mesophases in polymers has been already the subject of authored reviews in the past literature, for instance the cited article of Wunderlich and Grebowitz [2].

The above definitions given for the classes ii) and iii) of solid mesophases are sufficiently general to include a large number of disordered mesomorphic structures described for polymers so far. In fact, they are not based on the particular kind and amount of disorder present in the crystals but they rather address idealized structural models of disorder, where long-range order in the positioning, in three-, two- or one-dimension, of structural features, which are not necessarily point centered, is still present [13]. Not-point-centered structural features are, for instance, chain axes, centers of gravity of special groups of atoms along the chains, bundles of chains and so on. Within this framework, "condis crystals" [2] fall into one of the two classes ii) or iii) and there is no need for identifying a separate category of mesophases.

It is worth noting that the definition "condis crystals", in the sense given by Wunderlich [2], includes cases where the conformational disorder is essentially dynamic, and mainly has an entropic origin. In the following section, we will show that conformational disorder may also have a different origin because of the presence of constitutional and/or configurational irregularities along the chains.

The term "mesophase" also includes ordered liquids (nematic, smectic, cholesteric and discotics), which present long-range orientational order like in a solid, but positional disorder like in a liquid [2]. In these materials, large-scale molecular motion is possible, which is a characteristic of the liquid state rather than of the solid state. The term "liquid crystals" is conventionally used to address them. This sub-class of mesophases will not be treated in this context.

Let us point out that according to Wunderlich [2], "plastic crystals" may also be considered as mesophases. They are characterized by positional order but orientational disorder of the structural motif. Molecules of plastic crystals are generally close to spherical so that there is no high-energy barrier to their reorientation. Of course, the condition of a spherical shape of the molecules may not be fulfilled by the macromolecular chains of linear synthetic polymers, which generally crystallize in extended chains or helical conformations [2]. However, there is at least one case of crystals of macromolecules presenting orientational disorder of the structural motif as in plastic crystals.

This case has been recently described in the literature and is represented by the highly disordered "crystalline" form of alternated ethylene-norbornene (EN) copolymers [121]. The "crystallinity" of EN copolymers will be discussed in a following section, pointing out at the structural analogies with the conventional plastic crystals.

In the following sections we will analyze some examples of well-known polymeric solid mesophases, which may be classified as belonging to class ii) or iii). The structure of these mesophases has often been the subject of conflicting opinions and debates in the literature and their formation has often involved fundamental aspects of polymer science [122].

It is worth mentioning the hexagonal phase of polyethylene, stable at high pressures and temperatures [15–17, 123], which is considered the archetype mesophase, subject of large attention in the past and current literature, with large implications in the fundaments of polymer science [122–125]. Since its early identification, it has been argued that the formation of mesomorphic bundles of polyethylene chains plays an important role in the growth of extended chain crystals at elevated pressure [15–17, 123]. Later, much work of Keller's group has shown that the hexagonal phase can exist as a metastable, transient, highly mobile mesophase, at much lower pressure and temperature than reported earlier, i.e. in the stability region of the *PT* phase diagram of polyethylene of the usual orthorhombic form [124–126]. These studies drew attention to the possibility that the formation and growth of crystals requires the presence of the mobile phase under all circumstances, in the case of polyethylene in particular and of crystalline polymer in general [124–126].

Experiments generally indicate that the formation and growth of the lamellar crystallites is a multi-step process passing over intermediate states [127–131]. Recently, based on broad and detailed evidence from a large variety of experiments on several polymer systems carried out by many authors, a novel concept for understanding the crystallization from the melt of polymers has been developed by Strobl [132]. It is proposed that the initial step in the formation of crystalline lamellae always consists in the creation of a mesomorphic layer which spontaneously thickens, up to a critical value, where it solidifies through a cooperative structural transition. The transition would produce a granular crystalline layer, which transforms in the last step into homogeneous lamellar crystallites [132]. This issue still remains an open question [133–135].

In this review, the hexagonal phase of polyethylene stable at high pressure and temperature will be discussed, focusing on structural aspects, without further mention of collateral implications to its formation and stability.

4 Calculation Methods of the X-ray Diffraction Intensities from Disordered Model Structures

The presence of disorder in the crystals gives rise to diffuse scattering in a diffraction experiment, while few Bragg reflections may be still present in the diffraction patterns, depending on residual order.

The analysis of Bragg peak intensities allows to gain information concerning the average atomic positions and does not contain information regarding chemical or displacement correlations among neighboring structural motifs. The diffuse component of the diffraction patterns, instead, is strongly affected by the presence of these correlations and its analysis may help to elucidate the nature of disorder.

The quantitative treatment of diffuse scattering was pioneered by Warren [112, 136] and successively developed, in several of its many-fold aspects, by many authors [137–151]. A widely used approach consists of the derivation of analytical formulas for the calculation of X-ray diffraction intensity in terms of short-range chemical and/or displacement correlations associated with interatomic distances in the real space. In the hypothesis that short-range correlations are absent, disorder occurs at random, and this leads to a noticeable simplification in the formulas in use for the calculated scattered intensity.

In the case of crystalline polymers, the structural information deriving from use of X-ray diffraction techniques alone is generally poor, because the number of independent observed reflections is low and even lower in presence of structural disorder [152–154]. In the case of polymers, the structural analysis is an inductive process, rather than deductive as for conventional crystalline materials. A massive use of complementary computational techniques (geometrical and conformational analysis, simulation techniques) is thus required and, where possible, use of different experimental techniques, in addition to diffraction experiments, is useful in order to gain the highest number of *a priori* information [152]. The structural analysis in the case of solid mesophases of polymers is generally performed using these *a priori* information for building up structural models including various kinds of structural disorder, suitable for the calculation of X-ray diffraction intensities, which are compared with experimental diffraction data.

The collection of diffraction data of polymers in a solid mesophase for the structural analysis is generally performed on uniaxially oriented samples (fibers) [152–154]. Oriented fibers of high polymers in a mesomorphic form may be easily obtained by extrusion procedures from solution or melt and/or by cold stretching procedures.

From a first inspection of X-ray fiber diffraction patterns and with help of other experimental techniques (e.g. DSC analysis, solid-state NMR spectra, IR and Raman spectra), it is possible to deduce a series of structural information (chain periodicity, lateral packing of chains, chain conformation) and information concerning the possible kinds of disorder present in the structure. The most common kinds of disorder present in polymer crystals are disorder in the relative rotation of chains around their axes and in the relative translation along the chain axes [155, 156], conformational disorder [14, 58] and lattice distortions [59, 60]. Each kind of disorder affects in a different way and to a different extent the diffraction intensity distribution and therefore it is, generally, not easy to discern among the various possibilities and to establish the length scale range over which periodicity is still maintained. More quantitative information may be gained by comparison of experimental diffraction data with the diffraction intensity calculated for trial structures, modeling the system under analysis.

The non-perfect alignment of chain axes in the crystals with fiber axis results in an arcing of diffraction maxima. This effect may sometime complicate the interpretation of X-ray diffraction patterns but, usually, does not affect the interpretation of the experimental X-ray diffraction intensity distribution in terms of limit models of structural disorder.

Whenever necessary, the effect of the non-perfect alignment of chain axes in the crystals with fiber axis may be quantitatively analyzed in terms of moments of the orientation distribution of chain axes parallel to the fiber axis [157–159]. Ruland and Tompa in [158], showed that the scattering from a distribution of independent molecules is given by the convolution of the orientation distribution of molecular axes with the scattering of a single molecule. If both orientation distribution and molecular scattering are cylindrically symmetric, the resulting scattering has also cylindrical symmetry and all three functions may be expanded in a series of even-order Legendre polynomials, whose coefficients are mathematically related and correspond to the moments of orientation distribution or order parameters [157]. The most simple analysis is based on the fitting of the azimuthal profile of a single arbitrary reflection with a series expansion of Legendre polynomials [160]; this, for diffracting objects with a cylindrical symmetry (fibers), allows to access to the orientation distribution function of chain molecules.

The scattering function (I_{tot}) of a disordered crystal may be generally considered as given by the sum of two contributions, the intensity scattered from a hypothetical average lattice (I_{id}) and the diffuse scattering due to the presence of disorder (I_{diff}), according to Eq. 1:

$$I_{tot} = I_{id} + I_{diff} \,. \tag{1}$$

At a given reciprocal point $\boldsymbol{s} = \eta\boldsymbol{a}^* + \chi\boldsymbol{b}^* + \lambda\boldsymbol{c}^*$, with η, χ and λ real numbers, $\boldsymbol{a}^*$, $\boldsymbol{b}^*$ and $\boldsymbol{c}^*$ the reciprocal lattice parameters, the intensity scattered from an average lattice may be evaluated by Eq. 2:

$$I_{id}(\boldsymbol{s}) = L^2 \left\langle |F(\boldsymbol{s})| \right\rangle^2 \,. \tag{2}$$

where $\langle|F(s)|\rangle^2$ is the square of the Fourier transform modulus of the density function of a single structural motif (structure factor), averaged over all possible atomic positions due to the presence of disorder; for a periodic three-dimensional array of N_a, N_b and N_c structural motifs along a, $b$and c, respectively, L is the product of Laue functions along the a, b and c lattice directions. For a large periodic array, Eq. 2 is non zero only for $h=\eta$, $k=\chi$, $l=\lambda$ integer numbers. Therefore, Eq. 2 represents the Bragg contribution to the total scattered intensity.

For distortions of the first kind, neglecting thermal disorder and in absence of lattice disorder displacement, the diffuse part of the X-ray scattering (I_{diff}) is given by Eq. 3 (see for instance [152]):

$$I_{\text{diff}}(s)=\sum_{m=1}^{N}\sum_{n=1}^{N}\left\langle\Delta F_n\Delta F^*_{n+m}\right\rangle\exp(-2\pi\mathrm{i}s\cdot\boldsymbol{r}_m)\,. \tag{3}$$

where $\boldsymbol{r}_m$ is the separation distance between two structural motifs, $\Delta F_n=F_n-\langle F_n\rangle$ is the difference between the structure factor of a motif in the n-th lattice site and its average value and $\Delta F^*_{n+m}=F^*_{n+m}-\langle F^*_{n+m}\rangle$ is the complex conjugate of the corresponding quantity for the structural motif displaced m lattice sites from the n-th site. The terms $\langle\Delta F_n\Delta F^*_{n+m}\rangle$ represent the correlation functions of the fluctuations of scattering function. Equation 3 is a rapidly vanishing function as the distance between interfering motifs increases and describes a broadly distributed, smoothly varying intensity component.

In absence of any short-range correlation, Eq. 3 can be noticeably simplified, and the diffuse scattering may be calculated through Eq. 4 [152]:

$$I_{\text{diff}}(s)=N\left(\left\langle|F(s)|^2\right\rangle-\left\langle|F(s)|\right\rangle^2\right)\,. \tag{4}$$

where $\langle|F(s)|^2\rangle$ is the mean square modulus of F and N is the number of structural motifs in the crystal. Equation 3 is widely more general than Eq. 4, and allows to account in explicit for occurrence of eventual local correlations.

The Laue function in Eq. 2 for small size crystals, as in crystalline polymers, is characterized by spurious, undesirable maxima which, in a polycrystalline material are quite absent, owing to the large distribution of crystal dimensions. A Bernoulli-type distribution of the size of the crystallites may be in these cases assumed substituting the product of Laue functions in Eq. 2 with a product of functions of kind [150, 151]:

$$\frac{1-p_k^2}{1+p_k^2-2p_k\cos(2\pi s_k k)}\,. \tag{5}$$

where $1-p_k$ represents the probability of termination of a crystallite growth along the lattice direction k (= a, b or c) assumed, at each new step of growing process, independent of the length of the crystallite itself and s_k is the component of the scattering vector s along the reciprocal direction k^* ($= a^*, b^*$ or c^*). The probabilities p_k may be assumed to be related to the apparent length of the crystallites along the k lattice direction, L_k, through the Eq. 6 [96]:

$$p_k = \exp(-2k/L_k)\,. \tag{5}$$

In this case, the number of repeating structural motifs in the crystal, N in Eq. 4, should be replaced by the term $\langle N^2\rangle/\langle N\rangle$, i.e. the ratio of the mean square number and the mean number of structural motifs.

The presence of isotropic thermal disorder may be accounted for multiplying the Bragg contribution to the total intensity Eq. 2 by the Debye factor:

$$D = \exp\left(-8\pi^2 \langle u^2\rangle \sin^2\theta/\lambda^2\right). \tag{7}$$

where θ is the diffraction angle, λ the X-ray wavelength and $\langle u^2\rangle$ the mean square amplitude of atomic thermal vibrations.

Diffuse scattering arising from thermal displacement disorder, in the case of a crystal containing only atoms of one kind which vibrate independently about their average lattice sites, corresponds to a very simple formula, given by Eq. 8 [64, 152]:

$$I_{\text{thermal}}(s) = N_{\text{at}} f^2 (1 - D)\,. \tag{8}$$

where f is the atomic scattering factor and N_{at} the number of atoms in the crystal. This term contributes regardless of the presence of other kinds of disorder to the diffuse scattering expressed by Eq. 3 or Eq. 4 and should be added to Eq. 1 in order to account for diffuse scattering deriving from thermal disorder.

The presence of lattice distortions of second kind along a given lattice direction s_k may be accounted for replacing L^2 in Eq. 2 by the convolution product of L^2 with terms of kind:

$$\frac{1 - \exp\left(-4\pi^2\sigma^2 s_k^2\right)}{\left[1 - \exp\left(-2\pi^2\sigma^2 s_k^2\right)\right]^2 + 4\sin^2\left(\pi k s_k\right)\exp\left(-2\pi^2\sigma^2 s_k^2\right)} \tag{9}$$

This term is derived in the hypothesis that the distance between two neighboring points fluctuates around the average value k and its probability distribution is a Gaussian function with standard deviation σ [60].

Let us discuss some specific examples of lattice distortions of first kind.

As an example of uncorrelated disorder, it may be shown that diffuse scattering arising from a system containing two structural motifs which may substitute for each other at random in the lattice sites, without implying any distortion of lattice distances, corresponds to a very simple formula given by

Eq. 10 [63]:

$$I_{\text{diff}}(s) = \text{N}c_1c_2\,|F_1 - F_2|^2 \,. \tag{10}$$

where c_1 and c_2 are the relative fractions of the two mixed structural motifs 1 and 2 and F_1 and F_2 the corresponding form factors. Thus, the X-ray diffraction patterns of materials containing random substitutional disorder, in absence of any short-range correlation, would be characterized by the presence of halos over large regions of reciprocal space, along with the presence of sharp Bragg peaks.

Another simple, interesting case arises in polymers when the disordered structure does not present any short-range correlation in the relative height of chains along the z axis. Let $F(s)$ be the structure factor of a single periodic chain. In these cases, it may be shown that the mean value of $|F(s)|$ over the whole ensemble of chains is zero everywhere except along the equator and the meridian [64]. Thus, the Bragg contribution to the scattering intensity Eq. 2 reduces to zero on layer lines off the equator and off the meridian, whereas Eq. 4 further simplifies to the following equation:

$$I_{\text{diff}}(s) = \text{N}\left\langle |F(s)|^2 \right\rangle \,. \tag{11}$$

Therefore, in a fiber diffraction pattern, the presence of diffuse scattering localized along well-defined layer lines and absence of Bragg reflections off the equator generally indicate lack of any order in the relative height of the chains along the chain axes.

The effect of random and small angular displacements of chains around their axes and of small translation displacements of chains parallel to their axes on the diffraction intensity distribution in a fiber pattern may be treated as follows.

For fibers, the calculated intensity is conveniently expressed in terms of reciprocal cylindrical coordinates, ξ, ψ and ζ. Let us consider a generic aggregate of W parallel chains oriented with their axes (z) parallel to fiber axis. Without loss of generality, suppose that the chains are placed at the nodes of a large regular bidimensional lattice, each translated along z and rotated around z by z_w and φ_w, respectively. The structure factor of this chain aggregate is given $F(\xi, \psi, \zeta)$ [156]:

$$F(\xi, \psi, \zeta) = \sum_{w=1}^{W}\sum_{j=1}^{J} f_j \exp\left\{2\pi \mathrm{i}\left[\xi \varrho_j \cos(\varphi_j + \varphi_w - \psi) + \zeta(z_j + z_w)\right]\right\} \,. \tag{12}$$

where ϱ_j, φ_j and z_j are the cylindrical coordinates of j-th atom in a chain, referred to a local system of axes, with z coinciding with chain axis, f_j the atomic scattering factor and J the number of atoms in each chain. If $W = 1$, $F(\xi, \psi, \zeta)$ corresponds to the structure factor of the single chain.

Equation 12 may be expanded in series of Bessel functions of the first kind, obtaining the following expression [156]:

$$F(\xi,\psi,\zeta)=\sum_{w=1}^{W}\sum_{j=1}^{J}f_j\sum_{n=-\infty}^{+\infty}J_n(2\pi\varrho_j\xi)\times\exp\left\{i\left[n\left(\psi+\frac{\pi}{2}\right)-n(\varphi_j+\varphi_w)+2\pi\zeta(z_j+z_w)\right]\right\}. \tag{13}$$

with J_n the n order Bessel function. It is worth noting that for $W=1$, Eq. 13 corresponds to the Cochran, Crick and Vand formula for the calculation of the structure factor of a helical chain [161]. According to Cochran, Crick and Vand [161], for a commensurable helix including N residues in M turns, the order n of Bessel functions contributing to the X-ray diffraction intensity distribution along a given layer line with $l=\zeta/c$ (c, chain periodicity) may be selected solving the equation $l=nN+mM$ with m integer. More in general, for a helical chain characterized by pitch P and axial spacing p (unit height), the orders of Bessel functions entering Eq. 13 correspond to solution of equation $\zeta=n/P+m/p$ (generally used for incommensurate helices when M and N are not integer numbers) [161]. Of course, for chains which are not helical Eq. 13 is still valid, but the summation extends over all Bessel functions [161]. In practice, only low orders of Bessel functions need to be considered for the calculation of F with Eq. 13, because for a given argument, the value of a Bessel function rapidly decreases with increasing the order n.

The diffraction intensity in a fiber pattern $I(\xi,\zeta)$ corresponds to the cylindrically averaged product of $F(\xi,\psi,\zeta)$ (Eq. 13) with its complex conjugate $F^*(\xi,\psi,\zeta)$ as given by Eq. 14 [156]:

$$I(\xi,\zeta)=\sum_{n=-\infty}^{+\infty}\sum_{j=1}^{J}f_j^2J_n^2(2\pi\varrho_j\xi)(C_n+S_n)+\sum_{n=-\infty}^{+\infty}\sum_{j=1}^{J}\sum_{j'=1}^{J}f_jf_{j'}J_n(2\pi\varrho_j\xi)J_n(2\pi\varrho_{j'}\xi)(C_n+S_n)\times\cos\left[n(\varphi_j-\varphi_{j'})+2\pi\zeta(z_j-z_{j'})\right] \tag{14}$$

with:

$$C_n=\left[\sum_{w=1}^{W}\cos(2\pi\zeta z_w-n\varphi_w)\right]^2\quad S_n=\left[\sum_{w=1}^{W}\sin(2\pi\zeta z_w-n\varphi_w)\right]^2. \tag{15}$$

The average diffraction intensity of crystals including random disorder in rotational displacements of chains around their axes may be easily evaluated from Eq. 14, for any distribution of rotational displacements, by integrating the terms in the sums of Eq. 15 in the proper range, according to the assumed distribution. For instance, in the simple hypothesis that all $z_w=0$ and the chains are rotated by angles φ_w around the chain axis and these angles are uniformly distributed in the range 0–2π, all C_n and S_n terms in Eq. 15 are zero except for $n = 0$; in this latter case, indeed, $C_0 = W^2$. Displacement disorder of chains around their axes of a small angle φ_w symmetrically distributed around $\varphi_w = 0$ for all the chains, instead, implies $S_n = 0$ for all n, whereas the terms C_n can be approximated by a sort of an extended Debye-Waller factor, i.e. $C_n \approx W^2 \exp(-\, n^2 \langle \varphi^2 \rangle$ with $\langle \varphi^2 \rangle$ the average square value of φ_w [156].

Similarly, the average diffraction intensity of crystals including small translation displacement disorder of chains along their axes may be easily treated starting from Eq. 14, for any distribution of translation displacements, by integrating the terms in the sums of Eq. 15 in the proper range, according to the assumed distribution. For instance, in the simple hypothesis that all $\varphi_w = 0$, and the chains are translated at random by small distances in the direction of the chain axis and that these displacements are distributed symmetrically around $z = 0$, it may be shown that the averaging of Eq. 14 corresponds to setting $S_n = 0$ and $C_n = \exp(-\, 4\pi^2 \zeta^2 \langle z^2 \rangle)$ with $\langle z^2 \rangle$ the average square values of z_w [156].

The discussed examples of random rotational and translation displacement disorder of chains around their axes have important implications for the interpretation of the X-ray fiber diffraction patterns of poly(tetrafluoroethylene) oriented samples in the mesomorphic forms stable at high temperatures (Sect. 5.7).

Of course, the presence of non-random substitutional, translational or rotational displacement disorder, and hence the presence of short-range lateral correlations between neighboring chains, noticeably complicates the evaluation of the X-ray diffraction intensity distribution. Diffuse scattering in this case appears more localized; for instance, in the case of non-random substitutional disorder, for systems where two different structural motifs tend to alternate in neighboring sites, diffuse scattering appears rather peaked near superstructure positions in the diffraction patterns [112]. In these cases, Eq. 3 should be used instead of Eq. 4 for the calculation of diffuse scattering. Equation 3 is usually presented in the form [61, 112]:

$$I_{\text{diff}}(s) = \left\langle |\Delta F|^2 \right\rangle K(s)\,. \tag{16}$$

The function $K(s)$ characterizes the short-range order and it is defined by Eq. 17:

$$K(s) = \sum_{m=1}^{N} \frac{\left\langle \Delta F_n \Delta F^*_{n+m} \right\rangle}{\left\langle |\Delta F_n|^2 \right\rangle} \exp(-\, 2\pi i s \cdot r_m)\,. \tag{17}$$

In the case of absence of any short-range order, i.e. of uncorrelated disorder of the molecules, $K(s) = 1$.

The case of disordered structures where correlation must be taken into account has been treated by several authors [61, 137, 142, 147, 149–151]. For instance, Allegra [149] has developed a general approach for the calculation of diffracted intensity by monodimensionally disordered structures, in which different structural motifs follow each other along a given lattice direction according to different translation vectors with different probabilities according to a Markov statistics; this approach makes use of a matrix formalism which allows easily to account for any kind of short-range correlation in the succession of different motifs.

An alternative route for the calculation of X-ray diffraction intensity of a fiber, consists of developing the structure factor of a chain aggregate in a series of Bessel functions of zero order [152]; the cylindrically averaged diffraction intensity $I(\xi,\zeta) = \langle F(\xi,\psi,\zeta)F^*(\xi,\psi,\zeta)\rangle_\psi$ of a chain ensemble of M atoms with coordinates x, y and z referred to a common cartesian system of axes may be expressed in this case by Eq. 18:

$$I(\xi,\zeta) = \sum_{j=1}^{J}\sum_{j'=1}^{J} f_j f_{j'} J_0(2\pi\xi r_{ij}) \exp(2\pi\zeta z_{jj'}) . \tag{18}$$

where $r_{jj'}(= \left[(x_j - x_{j'})^2 - (y_j - y_{j'})^2\right]^{1/2})$ and $z_{jj'}$ $(= z_j - z_{j'})$ are the distances in the xy plane and along z, respectively, between atoms j and j'.

In the most crude approach, Eq. 18 may be used to evaluate the diffraction intensity of model structures containing specific kinds of disorder. The terms of summation rapidly fade away for large distances between interfering atoms. In some cases it may be convenient to multiply each term in the sums of Eq. 18 by paracrystalline disorder factors $g_{ij'}$ which reduce the interference between each couple of atoms according to the following equation [152, 162]:

$$g_{jj'} = \exp(2\pi^2\zeta^2 u_\zeta^2 z_{jj'}/\Delta_z) \exp(2\pi^2\xi^2 u_\xi^2 r_{jj'}/\Delta_r) . \tag{19}$$

According to Eq. 19, this reduction increases with increasing distance between the atoms; the terms u_ζ^2 and u_ξ^2 in the exponential functions represent the mean square displacements of the inter-atomic distances along z and in the xy plane, respectively, and are scaled by factors Δ_z and Δ_r. The quantities u_ζ^2/Δ_z and u_ξ^2/Δ_r are used as parameters in the calculations to be optimized in order to obtain the best agreement between calculated and experimental diffraction data.

Equation 18 is particularly useful for the calculation of the diffuse scattering of a mesomorphic aggregate of chains which does not present any correlation in the relative height of chains along z axis. In this case, in fact, we have shown that the scattering intensity of the whole bundle, off the equator, may be calculated by the average square of the modulus of the structure factor of the single chain, so that the sums in Eq. 18 can be extended to only the atoms of a single chain.

Termination effects due to the finite length of the model chain used in the calculations may be eliminated in the calculations, assuming, for instance, a Bernoulli type distribution of chain length, which corresponds to multiplying each term in the sums of Eq. 18 by factors of kind [163]:

$$t_{jj'} = (1 - 2p_c)^{|z_{jj'}|/\langle c\rangle} \tag{20}$$

with $1/p_c$ the average number of monomeric units in the chain, and $\langle c\rangle$ the average periodicity.

5 Solid Mesophases with Long-Range Positional Order in Three Dimensions of Not-Point-Centered Structural Features

5.1 1,4-tans-poly(1,3-butadiene)

The high-temperature crystalline form of 1,4-*trans*-poly(1,3-butadiene) was first described by Natta and Corradini [56] and then by Corradini [8] as a disordered crystalline phase characterized by conformational disorder of the chains.

1,4-*trans*-poly(1,3-butadiene), indeed, has two crystalline modifications. The crystalline form stable at room temperature transforms at 76 °C into a disordered modification [55]. The transition is reversible and the change of entropy involved at the transition (6.3 u.e./mole of mononomeric units) is more than twice larger than the entropy associated with the melting at higher temperatures (2.7 u.e./mole of mononomeric units) [57]. Such a large entropy change may be attributed to the conformational freedom gained by the chains at the crystal-crystal phase transition.

The crystalline phase stable at room temperature (form I) is characterized by a pseudo-hexagonal packing of chains at a distance of ≈ 4.55 Å; the chains have a **t***i* symmetry and periodicity $c = 4.85$ Å [56]. Actually, the symmetry is broken and a monoclinic unit cell has been suggested to explain the features of the X-ray fiber diffraction patterns ($a = 8.63$ Å, $b = 9.11$ Å, $c = 4.85$ Å, $\beta = 114°$, space group $P2_1/a$, four chains per unit cell) [164]. The chains have a regular conformation $(A^+ \; trans \; A^- T)_n$ shown in Fig. 3A in two projections parallel to the chain axis [8]. The backbone torsion angles adjacent to double bonds are in a skew conformation (*A* standing for anticlined) and alternate regularly along the chain with opposite signs (A^+ and A^-) whereas the backbone dihedral angles around $CH_2 - CH_2$ bonds are in the *trans* conformation (T). The chain periodicity would correspond to the experimental value of the chain axis ($c = 4.85$ Å) for $T = 180°$ and $A^+ = -A^- = 125°$ [8].

At 76 °C the low temperature form I transforms into the disordered crystalline form II, which melts at 146 °C [55, 56, 66].

The X-ray fiber diffraction patterns of form II show two equatorial reflections and a diffuse meridional 002 reflection [66, 165]. The equatorial reflections indicate a pseudo-hexagonal packing of chain axes with $a = 4.88$ Å and $c = 4.65$ Å [66]. During the transition from form I into form II, the c axis shrinks by 4% while the distance between the chain axes increases by 7.5% and the volume per chain in the crystalline state increases by 9% [8].

The high entropy change involved in the crystal-crystal phase transition and the features of the X-ray fiber diffraction patterns of form II could be accounted for assuming a disorder in the conformation of chains in form II [8]. Because of the low barrier energy which separates the intrinsic torsional energy minima around single bonds adjacent to the double bonds, A^+, A^- and C ($C = cis$) [65], chains of *trans*-1,4-poly(1,3-butadiene) present three non-equivalent, nearly isoenergetic conformational energy minima corresponding to the conformations $...A^{\pm}\,trans\,A^{\pm}T...$, $...A^{\pm}\,trans\,A^{\mp}T...$ and $...C\,trans\,A^{\pm}T...$ [166]. Therefore, a possible disorder in the conformation of 1,4-*trans*-poly(1,3-butadiene) may arise from a statistical succession of monomeric units which assume these energy minimum conformations [8, 58]. For instance, the random succession of sequences $...A^{\pm}\,trans\,A^{\pm}T...$ and $...C\,trans\,A^{\pm}T...$, characterized by the distance between the centers of consecutive CH = CH bonds of 4.84 and 4.48 Å, respectively, produce a contraction of the chain axis which matches the observed reduction from form I (4.83 Å) to form II (4.66 Å) for a probability of 25% that the single bonds adjacent to double bonds are in the *cis* conformation [166]. This model accounts for the X-ray diffraction data [66] and has been confirmed by the solid-state ^{13}C NMR analysis of form II [167]. The chains can be maintained straight, and hence the crystallinity is maintained, even with a statistical succession of structural units having widely different torsion angles, by allowing small deformation of bond angles [8].

Two possible models of the chains of 1,4-*trans*-poly(1,3-butadiene) in the mesomorphic form II are shown in Fig. 3B, as an example.

These disordered conformations not only account for the chain axis of form II but, on a semi-quantitative ground, also for the high value of the entropy change involved in the crystal-crystal phase transition between form I and form II and the low melting enthalpy of form II [8]. In fact, assuming equal statistical weights for the *cis* and *skew* conformations for the dihedral angles around the single bonds adjacent to double bonds, the number of rotational isomers for each monomeric unit accessible in form II is 9, resulting in a conformational entropy change for the crystal-crystal phase transition of 4.4 e.u./m.u.. This value, although it overestimates the entropic conformational contribution of the entropy change associated to this transition, is in good agreement with the calorimetric data. It accounts for more than one-half the experimental value of entropy change associated with the crystal-crystal phase transition of form I into form II, and for the low value of the melting entropy of 1,4-*trans*-poly(1,3-butadiene) [57, 168].

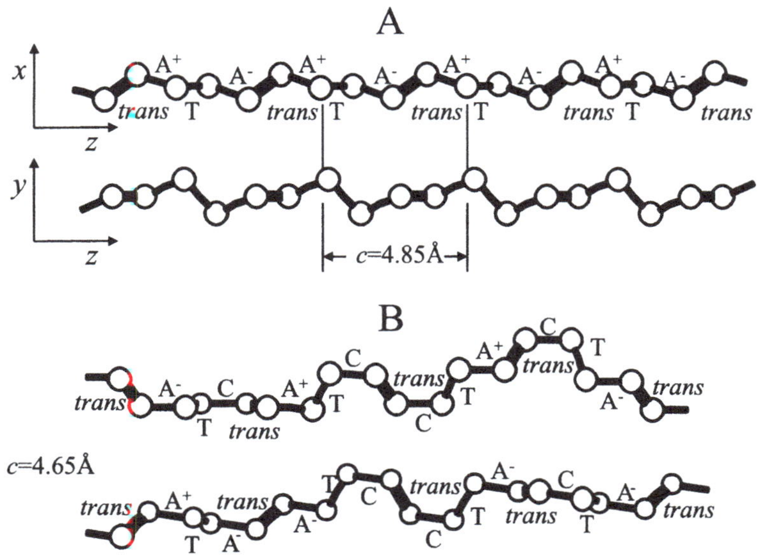

Fig. 3 (**A**) Chain conformation of 1,4-*trans*-poly(1,3-butadiene) in the crystalline form stable at low temperature (< 76 °C); chain periodicity $c = 4.85$ Å. (**B**) Two possible disordered chain conformations of 1,4-*trans*-poly(1,3-butadiene) in the crystalline form stable at $T > 76$ °C; average chain periodicity $c = 4.65$ Å. Symbols A^+, A^-,T and C stand for dihedral angles around C – C simple bonds close to + 120°, – 120°, 180° and 0°, respectively

The disordered form II of 1,4-*trans*-poly(1,3-butadiene) can be classified as a solid mesophase of type ii), where crystals have long-range order only in the position of not-point-centered structural features, that is, the chain axes. The two periodicities along *a* and *b* axes are sufficient to define a three-dimensional repetition, whereas the periodicity along the chain axis is defined by the average distance between the center of mass of consecutive monomeric units.

In addition, form II is a condis crystal in the original sense of Wunderlich [2], where cooperative motion between various conformational isomers is permitted, owing to the low interconversion barrier. A high mobility of chain stems in the crystalline regions of form II originating from such conformational motion, indeed, is also indicated by solid-state ^{1}H [169] and ^{13}C NMR analysis [167]. Probably, in the crystalline regions, the cooperative motion between various conformational isomers suggested by Corradini [8] is combined with chain rotation of the whole chain stems, as proposed by Iwayanagi and Miura in [169].

In conclusion, form II of 1,4-*trans*-poly(1,3-butadiene) represents a case of conformational isomerism of polymers in the crystalline state where different conformations occur at random within the same chain and long-range order is maintained among parallel chains [58].

5.2 Poly(ε-caprolactame) (nylon 6)

The β crystalline form of poly(ε-caprolactame) (nylon 6), obtained in melt spun fibers rapidly cooled to room temperature [24, 170] or by quenching the melt [23], is a solid mesophase, which, at variance with the form II of 1,4-*trans*-poly(1,3-butadiene), is metastable. The amount of disorder in the mesomorphic β form of nylon 6 is indeed related to the history of the sample, whereby non-equilibrium situations are kinetically frozen at room temperature; the β form transforms into the more stable α and/or γ forms by annealing at temperatures higher than room temperature and this transformation is irreversible [171].

As shown in Fig. 4, the X-ray diffraction patterns of nylon 6 fibers in the β form (Fig. 4A) present only two equatorial reflections (Fig. 4B) and several meridional reflections (up to the 7-th order, Fig. 4C) [163]. This indicates a pseudo-hexagonal unit cell with $a(=b) = 4.80$ Å and a mean periodicity along the chain axis equal to $c = 8.35$ Å. The substantial absence of diffraction off the equator and off the meridian indicates that a high degree of conformational disorder is present. From the meridional profile reported in Fig. 4C it is apparent that the half-height widths of the meridional reflections increase with increasing diffraction order, indicating paracrystalline distortions of the lattice parameter c.

The chain conformation for the ordered α [23] and γ [172] forms of nylon 6, as reported in the literature, correspond to chiral 2/1 helices with all coplanar amide groups. In both crystalline forms the chains are in nearly

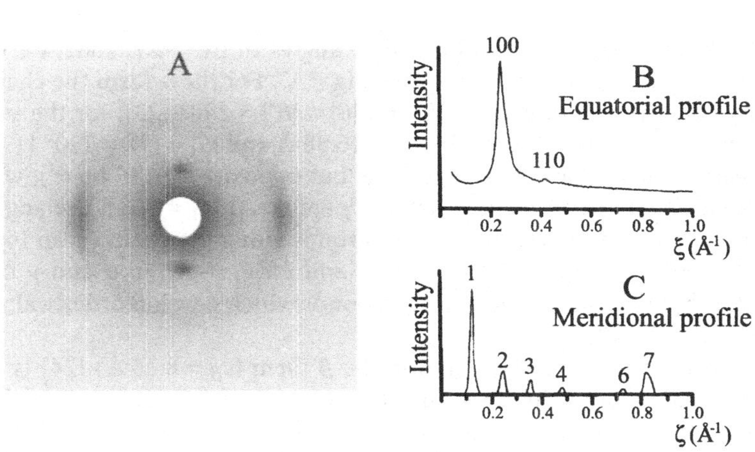

Fig. 4 (**A**) X-ray fiber diffraction pattern of an oriented sample of nylon 6 in the β form. (**B**) Equatorial profile; (**C**) Meridional profile. Pseudo-hexagonal unit cell, $a = b = 4.80$ Å, $\gamma = 120°$, average chain periodicity $c = 8.35$ Å (Reprinted with permission from [163]. Copyright 1997 by the American Chemical Society)

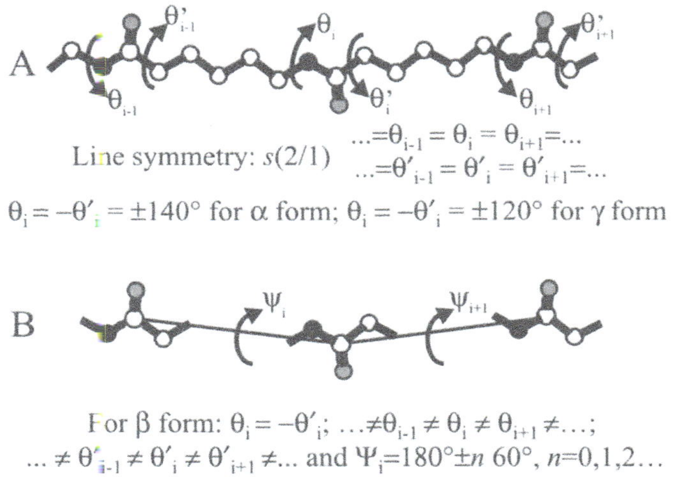

Fig. 5 (**A**) Schematic representation of nylon 6 chain. Symbols θ and θ' indicate the backbone dihedral angles in the repeating constitutional units adjacent to the amide groups. $\theta_i = \theta_{i+1}$ and θ_i=-θ'_i in the chains of α and γ forms in $s(2/1)$ extended conformation. The C – C bonds in the aliphatic chains are assumed to be close to 180°. (**B**) Symbols ψ_i identify the angles between the planes containing consecutive amide groups. The line symmetry $s(2/1)$ corresponds to all $\psi_i = 180°$, whatever the values of θ_i and θ'_i angles, with all coplanar amide groups. In the chains of β form the values of θ_i torsion angles are different, $\theta_i \neq \theta_{i+1}$, whereas still θ_i =– $\theta_{i'}$. Assuming θ (θ') variable, in absolute value, in the range 120–180°, large deviations of ψ_i from 180° would occur along the chain. Only conformations leading to $\psi_i = 180° \pm n60$, with n integer, are selected for nylon 6 in the β form

extended conformation with all dihedral angles in the *trans* state, but those adjacent to amide groups, θ and θ' (see Fig. 5A). For the α form the chain periodicity corresponds to $c = 17.2$ Å and $|\theta| = |\theta'| = 140°$ [23]; for the γ form the chain periodicity corresponds to $c = 16.88$ Å and $|\theta| = |\theta'| = 120°$ [172]. In both cases, θ and θ' have equal modulus but opposite sign (θ =– θ'); we also notice that under the restraints that all θ_i are equal ($\theta_i = \theta_{i+1}$), the angle between the planes of consecutive amide groups along the chain (ψ in Fig. 5B) is equal to 180°, whatever the value of θ and for θ_i =– θ'_i. In α and γ crystal forms the chains form strong hydrogen bonds which develop orderly along all parallel lines [23, 172].

The chain periodicity of nylon 6 in the β form ($c_\beta = 8.35$ Å) [24] is about one-half that of nylon 6 in the γ form ($c_\gamma = 16.9$ Å) [172] and slightly shorter than one-half the chain periodicity of nylon 6 in the α form ($c_\alpha = 17.2$ Å) [23]. This indicates that the chains in the mesomorphic β form present highly disordered, nearly extended conformations, with a periodicity along the chain axis defined by the average distance along the c axis between the center of mass of consecutive monomeric units [24]. A structural model of β form of nylon 6 should, therefore, include conformationally disordered chains in extended conformations packed in a pseudo-hexagonal array, and should ac-

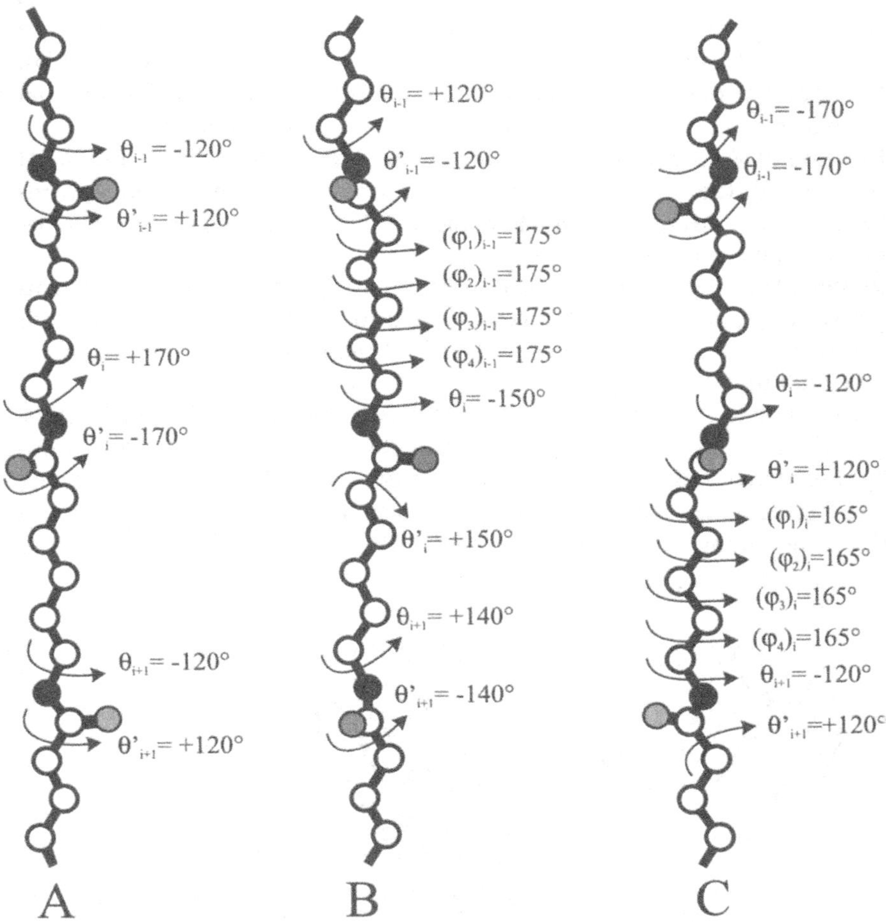

Fig. 6 Disordered conformations of nylon 6 chains of low internal energy suitable for the β form, built according to the simplifying assumptions of [163]. The average chain periodicity is $c = 8.35$ Å and the dihedral angles between consecutive amide groups along the chain (Ψ) are close to $\pm 120°$ or $180°$. According to [163], values of ψ close to ± 120 may be obtained for instance (**A**) under the condition that all backbone dihedral angles of a aliphatic chain (φ) comprised between two consecutive amide groups be equal to $180°$, when $(\theta')_i + (\theta)_{i+1} = \pm 70$; (**B**) under the condition that $(\theta')_{i-1} = -120°$ and $\theta_i = -150°$ when all $(\varphi)_i = 175°$; (**C**) under the condition that $(\theta')_i = -(\theta)_{i+1}$ when all $(\varphi)_{i+1} = 165°$. The straightness and high extension of the chain is ensured, retaining the condition $\theta_i = -\theta'_i$ for any i value

count for the experimental observation that in the β form hydrogen bonds are formed almost 100% as in the more stable α and γ forms [173, 174].

Possible chain models for the β form of nylon 6 in extended conformations including large amounts of conformational disorder were proposed in [163]. They were derived from those of nylon 6 in the α and γ forms, considering that the dihedral angles adjacent to amide groups, θ and θ', have low torsional barrier energy and are variable (in absolute value) in the range 120–180°, in agreement with structural data of a variety of amide compounds [174, 175]. The conditions of $s(2/1)$ symmetry that all the dihedral angles $\theta_i(\theta_i')$ are equal was then relaxed and only the condition $\theta_i =- \theta_i'$ was retained (see Fig. 5). The value of ψ_i depends on the value of the dihedral angles θ_{i-1}' and θ_i. At variance with the case of the ideal ordered structures of the α and γ forms of nylon 6, the disordered model chains of β form does not present any defined chirality and the $C=O$ bonds point along any direction cylindrically distributed around the chain axis. Restricting the conformations of the chains to those with $\psi_i = 180 \pm n60°$ (n integer), low energy extended conformations with average periodicity c close to 8.35 Å of the kind shown in Fig. 6, were easily built up [163].

According to [163], a possible supra molecular aggregate of nylon 6 chains, in the β form, is drawn in Fig. 7, as an example. The chains are conformationally disordered and have average periodicity $c = 8.35$ Å. The chain axes are placed at the nodes of a bidimensional hexagonal lattice. The hydrogen bonds are formed along lines parallel to [100], [010] and [1$\bar{1}$0] lattice directions;

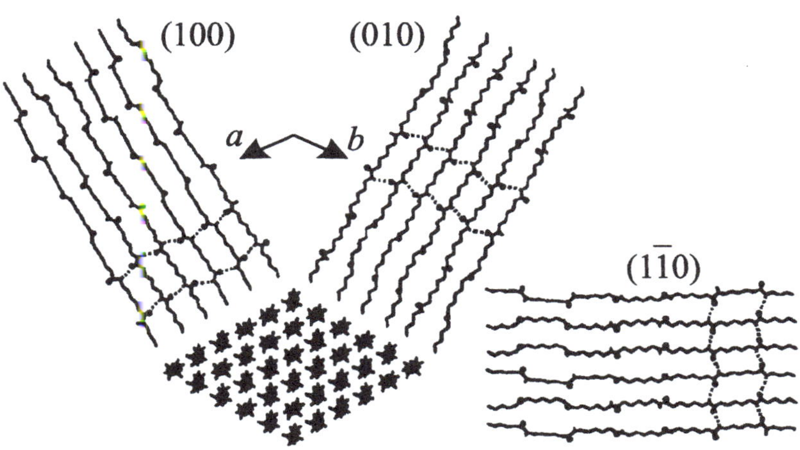

Fig. 7 Possible packing model for nylon 6 in the β mesomorphic form. The chain axes are placed at the nodes of a bidimensional hexagonal lattice. The concomitant formation of H-bonds along lines parallel to [100], [010] and [1$\bar{1}$0] lattice direction is indicated. (Reprinted with permission from [163]. Copyright 1997 by the American Chemical Society)

formation of hydrogen bonds is the driving force which induces neighboring chains within the small mesomorphic aggregates to changes in conformation so that nearly 100% of hydrogen bonds are formed [163]. As a consequence, since the amide groups lie all at nearly the same height along z, the lines of hydrogen bonds lie in layers perpendicular to the chain axes and have the same direction within each layer; however, lines of H-bonds in consecutive layers along z may be not parallel but rotated at chance by + 120 or – 120°.

In a recent paper, Li and Goddard III in [176] proposed that similar mesomorphic aggregates of nylon 6 chains are formed, as low energy intermediate structures, during the transformation between the α and γ forms.

The X-ray diffraction intensity calculated in [163] on such disordered arrays along the $\zeta = l/c$ reciprocal line (l a continuous variable) is in good agreement with the experimental X-ray diffraction profile along the meridian of Fig. 4C; small lattice distortions of the second kind along the c axis were introduced in the calculations, in order to account for the experimental observation that the half-height widths of the meridional reflections increase with increasing diffraction order (see Fig. 4C). The best agreement with the

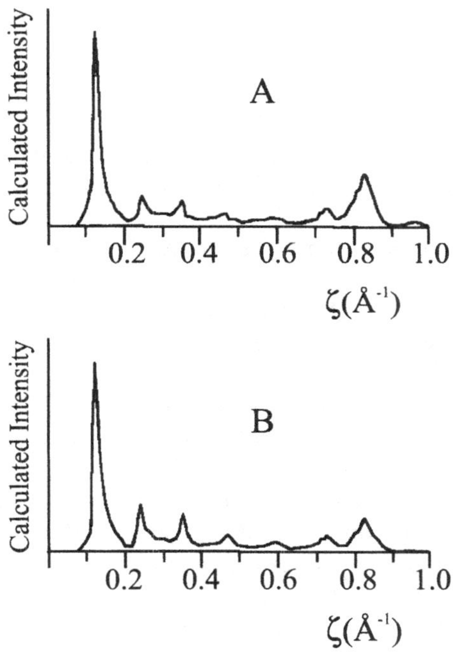

Fig. 8 (**A**) Calculated X-ray diffraction profiles along the meridian for model structures of β form of nylon 6 when only conformational disorder is present and the chain are all up or all down (isoclined), and (**B**) when conformational disorder and disorder in the substitution of up/down chains in the lattice positions are both present (Reprinted with permission from [163]. Copyright 1997 by the American Chemical Society)

experimental pattern was obtained introducing a further kind of disorder, arising from the random substitution in the lattice positions of β form of nylon 6, of chains with opposite orientation, i.e. with the $-\mathrm{NC}(=\mathrm{O})\mathrm{O}-$ amide groups pointing at chance either along a positive (up) direction parallel to the chain axis or toward the opposite (down) direction [163]. This is shown in Fig. 8, where the calculated X-ray diffraction profile along the meridian is reported in the case of model structures of nylon 6 in the β form including both conformational disorder and up/down substitutional disorder [163].

In conclusion, the β form of nylon 6 corresponds to small mesomorphic aggregates of chains of class ii) where order is maintained for structural features which are not point centered, on various length scales: long-range order is connected with the maintenance of the parallelism and of the pseudo-hexagonal arrangement of the chain axes (the size of the mesomorphic bundles perpendicular to chain axes is higher than the size parallel to chain axes); an intermediate range of order may be referred to the parallel arrangement of lines of hydrogen bonds within layers normal to the chain axes; an even shorter range of order refers to the piling of such layers along *c*, the size of the mesomorphic aggregates parallel to the chain axes spanning a few number (5 on average) periodicities parallel to the *c*axis.

5.3 Poly(acrylonitrile)

Poly(acrylonitrile) (PAN) is a crystalline polymer despite the irregular chain configuration: it is produced via radical polymerization and it is essentially atactic [39, 177].

The crystalline form of PAN may be classified as a solid mesophase of type ii): the PAN crystals or paracrystals have long-range positional order only for not-point-centered structural feature, that is the chain axes, for which the two $a = b$ periodicities are sufficient to define a three-dimensional repetition of the position of the chain axes [97].

The conformation of the stereo-irregular chains of PAN, which are able to crystallize in a pseudo-hexagonal form has been a matter of intense debates in the literature for years [39–48, 95–97, 177].

The X-ray diffraction pattern of a uniaxially oriented fiber of PAN is shown in Fig. 9 [97]. It presents only two sharp Bragg reflections on the equator (Fig. 9B), indicating a nearly perfect hexagonal arrangement of the chain axes in the a–b plane placed at distance of 6 Å [40, 48, 96, 97, 177] and diffuse halos off the equatorial line. Two halos along the meridian, centered at $\zeta \approx 0.40$ and 0.80 Å^{-1} (Fig. 9C), and a characteristic off meridian diffuse scattering, with cylindrical coordinates in the reciprocal space equal to $\xi = 0.15$ Å^{-1} and $\zeta = 0.25$ Å^{-1} (Fig. 9A), are present [95–97]. The position of the two broad maxima on the meridian suggests a value close to 2.5 Å for the average chain periodicity.

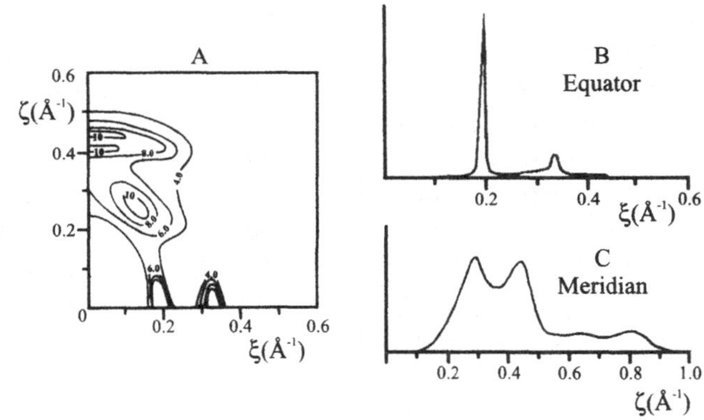

Fig. 9 Experimental X-ray diffraction intensity distribution of a PAN fiber (**A**) in the ξ, ζ reciprocal plane, (**B**) along the equator (for $\zeta = 0$) and (**C**) along the meridian (for $\xi = 0$). (**A**) : Contour lines with constant $I(\xi, \zeta)$ values at regular intervals of two units (arbitrary scale). (Reprinted with permission from [97]. Copyright 1996 by the American Chemical Society)

Consistent with the presence of the diffuse halo in the quadrant of X-ray diffraction patterns at $\xi = 0.15$ Å^{-1}, $\zeta = 0.25$ Å^{-1}, the chain periodicity of 2.5 Å was taken as an indication that the pseudo-hexagonal crystalline form of PAN is formed by elongated portions of chains with a prevailing syndiotactic configuration in a nearly *trans*-planar conformation [39–48]. In fact, fully extended conformations are at low cost of internal energy only for fully syndiotactic sequences, since *meso* sequences adopting a *trans*-planar conformation would lead to unfavourable dipole-dipole interactions between adjacent – CN groups [178–180]. However, according to X-ray diffraction data, assuming a packing of syndiotactic chains in a hexagonal unit cell with parameters $a = 6$ Å and $c = 5.1$ Å, (two monomeric units/unit cell), the theoretical density would be calculated equal to 1.11 g/cm^3, which is considerably lower than experimental density values of PAN indicated in the literature, in the range 1.17–1.22 g/cm^3 [181–183]. In the hypothesis of a hexagonal packing of chains in a unit cell with $a = 6$ Å, these values of density suggest a value of the chain axis lower than 2.5 Å and, hence, a conformation of the chain not exactly *trans*-planar.

Moreover, the hypothesis that only syndiotactic sequences in a fully planar conformation would crystallize is inconsistent with the distribution of stereosequences in PAN samples, which is essentially that expected in the case of a Bernoulli type statistics with equal *m* and *r* diads concentrations [184–186].

On the other hand, considering the characteristic X-ray diffraction patterns of PAN fibers, Bohn et al. [40] suggested that the pseudo-hexagonal form of PAN presents only lateral order in the packing of chain axes, and no longitudinal order parallel to the chain axes.

Lindenmeyer and Hosemann [48] analyzed the electron diffraction patterns of single crystals of PAN in terms of the theory of paracrystals, giving a plausible explanation for the absence or extreme weakness of high-order reflections. They argued that large fluctuations of the lattice vector parameter parallel to chain axes, corresponding to distortions of the second type (or paracrystalline distortions) along this lattice direction, may account for the experimental diffraction intensity distribution along the meridian. It was concluded that consistent with the paracrystal model, an approximate average periodicity close to 2.3–2.4 Å could be assumed for PAN atactic sequences in the crystalline state, achieving a better agreement with density measurements [48].

To reconcile the results of the X-ray diffraction studies with the tacticity [184–186] and density measurements [181–183], Liu and Ruland [96] suggested that short isotactic sequences could easily adopt a nearly zigzag planar conformation, whereas longer isotactic sequences would reduce the internal energy of the chain by formation of kinks. Each kink would be pinned to configurational *m* sequences of monomeric units embedded in syndiotactic sequences in a nearly *trans*-planar conformation; models of a kink are shown in Fig. 10, as an example: the dihedral angles close to *m* diads assume a TG^-TG^+ conformation. These kind of kinks would shorten the chains while maintaining the *trans*-planar portions of chain parallel (but not coaxial) on

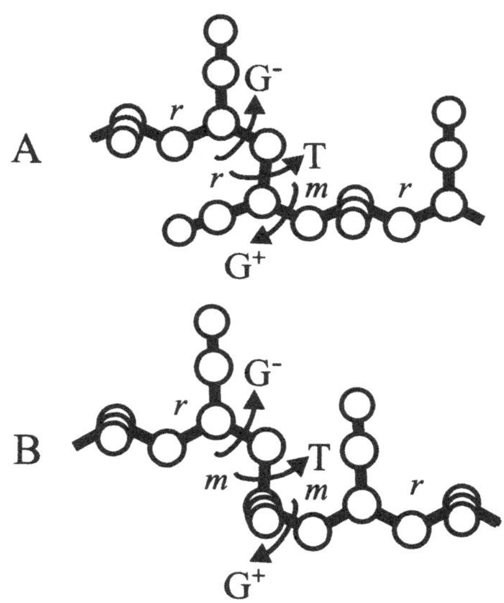

Fig. 10 Kink models for chain stretches of PAN including *m* configurational diads in a prevailing syndiotactic configuration. Kinks with conformational sequence G^-TG^+T shorten the chains while maintaining the *trans*-planar portions of chain parallel, but not coaxial, on both sides of the defect

both sides of the defect. The presence of about 1 kink per 10 monomeric units would account for the values of the experimental density [96].

In [97], an alternative model not implying kink formation was proposed based on the results of conformational energy calculations and geometrical analysis. The conformational space was sampled for short atactic portions of PAN chain containing *m* isolated diads and *mm* isolated triads in a syndiotactic configurational environment, with the aim to find extended conformations of minimum energy suitable for PAN in the crystalline state.

Although the analysis of [97] was not exhaustive, it succeeded in identifying extended conformations of minimum energy for stereoirregular model chains of PAN in which the torsion angles are close to 180° in *racemo* diads and may deviate significantly from 180° in *meso* diads. As an example, locally extended stretches of *rmr* tetrads and *rmmr* pentads in a minimum energy conformation are shown in Fig. 11.

According to the results of [97], nearly *trans*-planar conformations for configurationally irregular sequences are energetically feasible, with deviations of torsion angles of ±10 from 180° which alleviate unfavorable electrostatic interactions between adjacent nitrile groups [178–180]. These nearly *trans*-planar conformers correspond to a small lateral encumbrance of – CN groups and present an average chain periodicity (average length of the projections along the chain axis of the vectors connecting the center of mass of consecutive monomeric units) close to 2.5 Å (Fig. 11A and A'). This value is still in a bad agreement with indications deriving from the values of the experimental density.

Extended chain conformations including torsion angles in a *gauche* state for bonds around *m* diads are also feasible at low cost of conformational energy (Fig. 11B–C and Fig. 11B'–G'). The presence of *gauche* bonds introduces some waviness in the chain, which can be easily tolerated in the lattice while maintaining the chain straight and with small lateral encumbrance; *gauche* bonds, at the same time, shorten somewhat the mean chain periodicity of nearly *trans*-planar chain, from 2.5 Å to 2.3–2.4 Å, in a better agreement with density values [97]. In addition, the presence of *gauche* bonds in crystallizable portions of atactic chains, would rotate the mean plane of the backbone chain by nearly 120°, with the result that – CN side groups would project perpendicularly to the chain axis along directions placed at ≈ 120° each other.

In [97], the calculated Fourier transform of the disordered models of the isolated PAN chain of Fig. 11 including ...*rmr*... and ...*rmmr*... configurational sequences in different low minimum energy conformations, were compared to the experimental diffraction data. In the hypothesis that no long-range order in the relative shift of the chains along *z* would be present in the mesomorphic crystals of PAN and that the relative orientation of the chains around the chain axes would be related only on a local scale, the distribution of the diffracted intensity in the reciprocal space off the equator can be accounted for by the Fourier transform of the isolated chain.

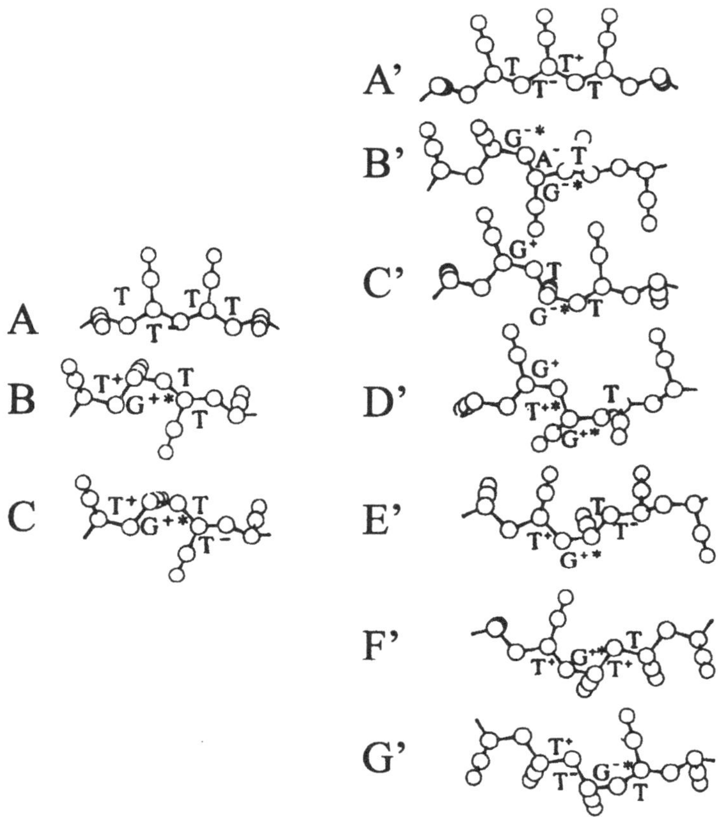

Fig. 11 Side views of chain stretches of PAN, with an isolated *m* diad (A–C) and an isolated *mm* triad (A'–G') between *r* diads in minimum energy conformations. Symbols G, A and T stand for torsion angles deviating up to ±30° from *gauche*, *anticlinal* and *trans* states, respectively. Starred A, G or T symbols indicate that, for the considered minimum energy conformation, the sign of the value of the dihedral angle is opposite with respect to the configurational sign (+) or (–) of the corresponding bond [65]. The inherent asymmetry of skeletal bonds is assigned according to cyclic order of the three groups attached to the tertiary carbon atoms, following definitions given in [65]. In the models B,C, B'-G', the average chain periodicity is ≈ 2.3–2.4 Å (Reprinted with permission from [97]. Copyright 1996 by the American Chemical Society)

The comparison between the calculated and experimental diffraction data indicated that the paracrystalline pseudo-hexagonal phase of PAN is formed by short and straight portions of chains with atactic configuration [97]. Model chains in a nearly *trans*-planar conformation for atactic PAN, do not account for the off-equatorial diffuse scattering in the X-ray diffraction patterns. A better agreement with the experimental diffraction data was obtained for atactic PAN chains containing C – C bonds close to *m* diads in *gauche* conformation. As an example, the calculated X-ray diffraction intensity in the ξ,

ζ reciprocal plane and along the meridian, for models of chains of Fig. 11 including ...*rmr*... and ...*rmmr*... sequences in different low minimum energy conformation, which gives a good qualitative agreement with the experimental X-ray fiber diffraction pattern of PAN of Fig. 9 is presented in Fig. 12. The configurationally disordered model chains of PAN used in the calculations of Fig. 12 includes up to 10% of dihedral angles close to *m* diads in the *gauche* conformation. These conformations present average chain periodicity close to 2.4 Å, shorter then 2.5 Å as required by density data and account for the position and the broadness of the meridional diffraction maxima. In addition, also the position and the intensity of the diffuse halo centered at

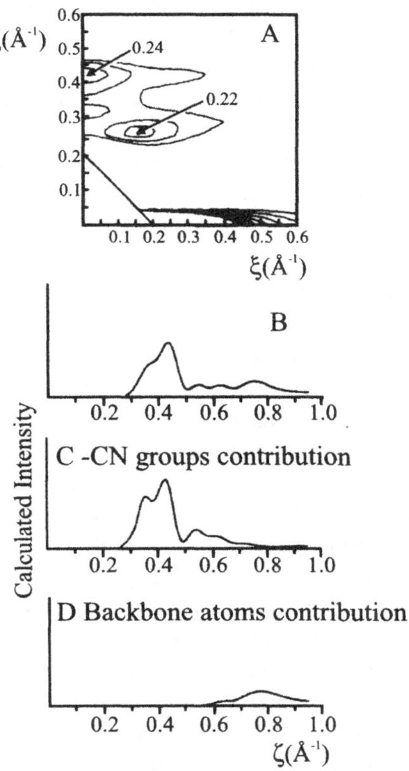

Fig. 12 (**A**) Calculated diffraction intensity I_c for disordered models of the isolated chain of PAN with atactic configuration including ...*rmr*... and ...*rmmr*... sequences in different low minimum energy conformations of Fig. 11, averaged over different chain models, as a function of the reciprocal cylindrical coordinates ξ and ζ, and (**B**)–(**D**) the corresponding profiles along the meridian. (**C**) The scattering intensity along the meridian arising from – CN groups alone and (**D**) from the backbone carbon atoms alone, are shown. In (**A**), the contour lines correspond to 0.05, 0.1, 0.15 and 0.20 arbitrary units (Reprinted with permission from [97]. Copyright 1996 by the American Chemical Society)

$\xi = 0.15$ Å^{-1} and $\zeta = 0.25$ Å^{-1} is well reproduced (Fig. 12A), indicating that a prevailing syndiotactic configuration is not necessary to account for the presence of this halo in the experimental pattern of Fig. 9A [97].

It is worth considering that in [97] an explanation of the origin of the diffraction maxima along the meridian at $\zeta \approx 0.40$ and 0.80 Å^{-1} is provided, consistent with the paracrystalline model proposed by Lyndenmeyer and Hosemann [48] for PAN. The halo at $\zeta \approx 0.80$ is not necessarily the second-order diffraction of the maximum at $\zeta \approx 0.40$ Å^{-1}, since it is apparent from calculations of Fig. 12 that these two meridional maxima may originate from different contributions. The maximum at $\zeta \approx 0.40$ Å^{-1} arises from the average periodicity of lateral – CN groups alone (Fig. 12C), whereas the maximum at $\zeta = 0.80$ Å^{-1} arises from the contribution of only the backbone carbon atoms (Fig. 12D).

In conclusion, the pseudo-hexagonal form of PAN may be described in terms of limit models of disorder implying the following features:

1. Maintenance of long-range positional order only as far as the positioning of chain axes; the chain axes are at the nodes of a bi-dimensional hexagonal lattice with $a = b = 6$ Å, and lateral dimensions covering several tens of angstroms.
2. The chains are almost extended, yet configurationally irregular and conformationally disordered. The chain conformation close to *racemo* diads is nearly *trans* whereas the backbone bonds close to *meso* diads may deviate largely from 180°, adopting a *gauche* conformation. The presence of *gauche* bonds shortens the mean repetition period from $c = 2.5$ to 2.4 Å/monomeric unit, consistent with density measurements. Corresponding to a *gauche* bond the direction of the pendant – CN groups results rotated by 120° with respect to the direction of the adjacent groups.
3. While there is long-range order in the positions of the chain axes, there is only a short-range order in the relative height of the chains along z. A local orientational order of – CN groups possibly directed along the lattice directions a, b, and $-(a + b)$, is also present.
4. The presence of configurational disorder in the crystalline state of atactic PAN necessarily implies the presence of conformational disorder to alleviate intramolecular strains. The conformational disorder, in turn, yields small deviations from the perfect straightness of crystalline portions of the chains, and thus some waviness, which comply more easily with the local packing of – CN groups.

Recent developments of solid-state NMR techniques have allowed a direct and detailed analysis of the local structure of PAN in the solid state using homonuclear two-dimensional (2D) double quantum (DQ) (DOQSY) [187, 188] and ^{2}H-^{13}C 2D heteronuclear multiple quantum (MQ) correlation (HMQC) [189] solid-state NMR spectroscopy. This method was applied to

PAN samples with chains ^2H and/or ^{13}C labeled at suitable positions, in order to selectively enhance NMR signals of atoms in *m* and *r* diads.

The results of this analysis basically confirm the model proposed in [97], that the chains in crystalline PAN are atactic, with conformations characterized by bonds close to *r* diads in *trans*-planar conformation slightly deviated from 180° and by the presence of bonds close to *m* diads in *gauche* conformation [188, 189]. According to solid-state NMR analysis of [187], the *trans-gauche* ratio is close to 90 : 10 in atactic PAN. Semiquantitative information on intermolecular alignment was also obtained by detailed analyses of 2D DOQSY spectra in [189], indicating that – CN side groups are mostly oriented along the crystalline lattice directions *a*, *b* and – (*a* + *b*) of the hexagonal lattice as in the structural model proposed in [97].

5.4 Ethylene-Propylene Random Copolymers

Random copolymers of ethylene with propylene (EP), with ethylene content between 80 and 40 mol % are amorphous materials at room temperature [30, 31] which, despite the irregular constitution, are able to crystallize at low temperatures or by stretching at room temperature [30–38, 190–193]. In EP copolymers, propylene units are included in the crystalline lattice of the orthorhombic form of polyethylene (PE, lattice parameters $a = 7.42$, $b = 4.95$, c (chain axis) = 2.54 Å [194]), inducing large disorder and decrease of degree of crystallinity.

Accurate X-ray diffraction measurements indicate that the dimension of the *a* axis of the unit cell of PE increases almost linearly with increasing propylene content, whereas the *b* and *c* axes practically retain the dimensions found in the PE homopolymer. For propylene content around 25% the a/b ratio approaches $3^{1/2}$ and the unit cell becomes pseudo-hexagonal [30, 32–38].

In particular, the X-ray diffraction pattern of a stretched sample of an ethylene-propylene terpolymer with a low amount of diene (< 2 mol %, EPDM), containing 75 mol % ethylene could be interpreted in terms of an orthorhombic unit cell with parameters $a_0 = 8.66$, $b_0 = 5.0$, c_0(chain axis)= 2.54 Å (subscript "o" standing for orthorhombic); the unit cell is pseudo-hexagonal with $a_0/b_0 = 3^{1/2}$ [30]. The relationship between orthorhombic (a_0, b_0) and hexagonal (a_h, b_h) unit cell parameters is illustrated in Fig. 13; for the hexagonal unit cell $a_h = b_h = 5$ Å whereas the c_h axis coincides with the orthorhomic c_0 parameter (subscript "h" standing for hexagonal)

The X-ray fiber diffraction pattern of a stretched sample of an EP copolymer in the pseudo-hexagonal form is shown in Fig. 14 [195]. The diffraction intensity is mainly concentrated along the equatorial layer line, indicating long-range order in the pseudo-hexagonal arrangement of chain axes. More-

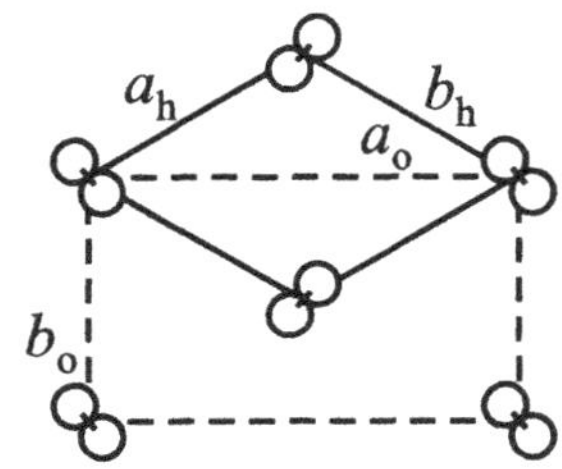

Fig. 13 Relationship between orthorhombic (a_o, b_o) and hexagonal (a_h, b_h) unit cell parameters in the structure of polyethylene. EP chains for propylene content close to 25%, pack in a pseudo-hexagonal unit cell, with orthorhombic lattice parameters in the ratio $a_o/b_o = 3^{1/2}$

over, the presence of broad peaks on well-defined layer lines indicates that long-range order is maintained within each chain, with a regular *trans*-planar conformation. The diffuse nature of the diffraction halos along the layer lines indicates that a large amount of disorder is present in the pseudo-hexagonal form of EP copolymers and that the coherent length of the ordered bundles of chains should be of the order of a few tens angstroms.

In [195] accurate X-ray diffraction measurements and comparison with Fourier transform calculations on model structures, has allowed to clarify several aspects concerning the nature of disorder which characterize the pseudo-hexagonal form of ethylene-propylene copolymers.

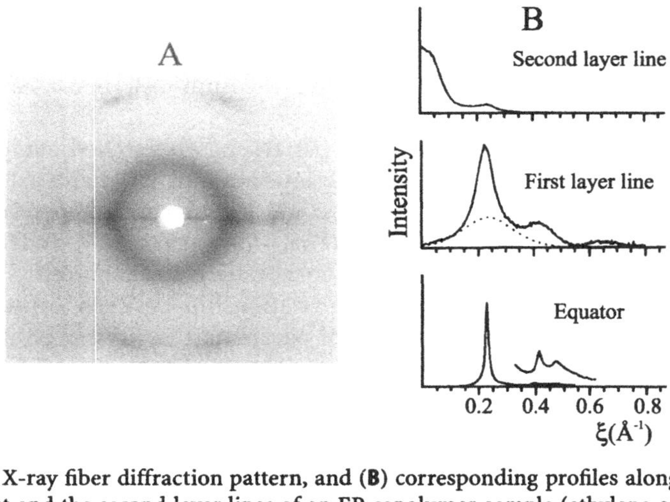

Fig. 14 (**A**) X-ray fiber diffraction pattern, and (**B**) corresponding profiles along the equator, the first and the second layer lines of an EP copolymer sample (ethylene content 75% mol) stretched at room temperature (elongation ratio 750%). (Reprinted with permission from [195]. Copyright 1996 by the American Chemical Society)

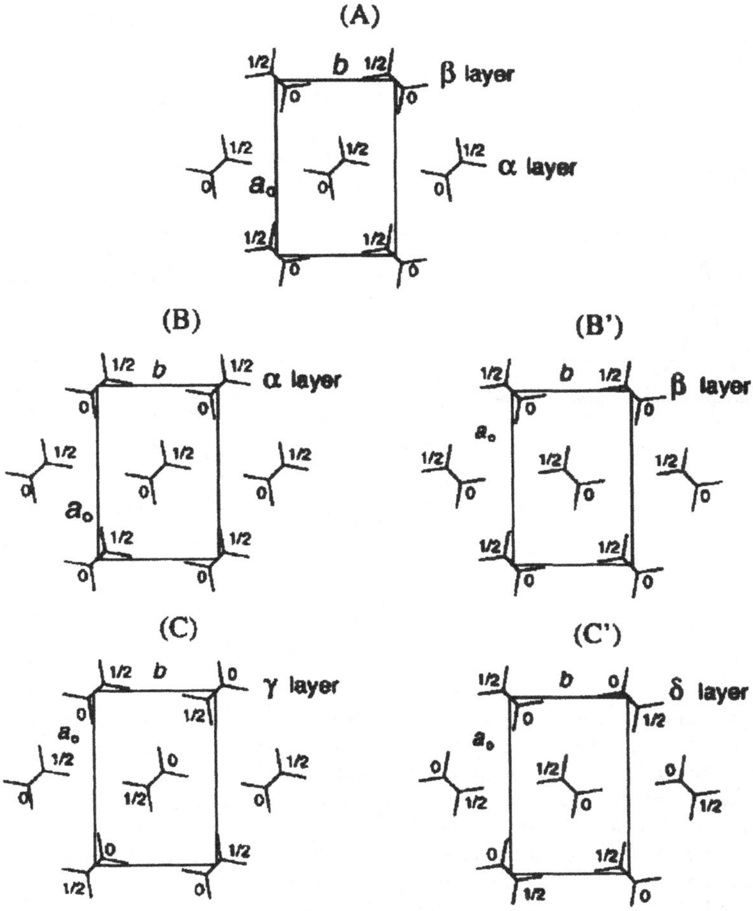

Fig. 15 Projections in the *ab* plane of the limit-ordered (**A**) orthorhombic, (**B,B′**) monoclinic and (**C,C′**) triclinic ideal model structures considered in the Fourier transform calculations for the pseudo-hexagonal form of EP copolymers in [195]. The numbers indicate the fractional *z* coordinate of the backbone carbon atoms. α, β, γ and δ layers indicate different kinds of $b_o c_o$ layer of chains piled along the a_o lattice direction. (Reprinted with permission from [195]. Copyright 1996 by the American Chemical Society)

Three limit-ordered models, shown in Fig. 15, were considered as possible ideal arrangements of EP chains in the mesomorphic bundles. In Fig. 15A and Fig. 15B,B' the chains are arranged as in the orthorhombic [194] and monoclinic [196] polymorphs of PE, respectively. In Fig. 15C,C' the chains are arranged as in triclinic form of long chains paraffins [197]. These models were chosen as reference, "ideal" structures, where different kinds of disorder were introduced, in order to better understand their influence on the cal-

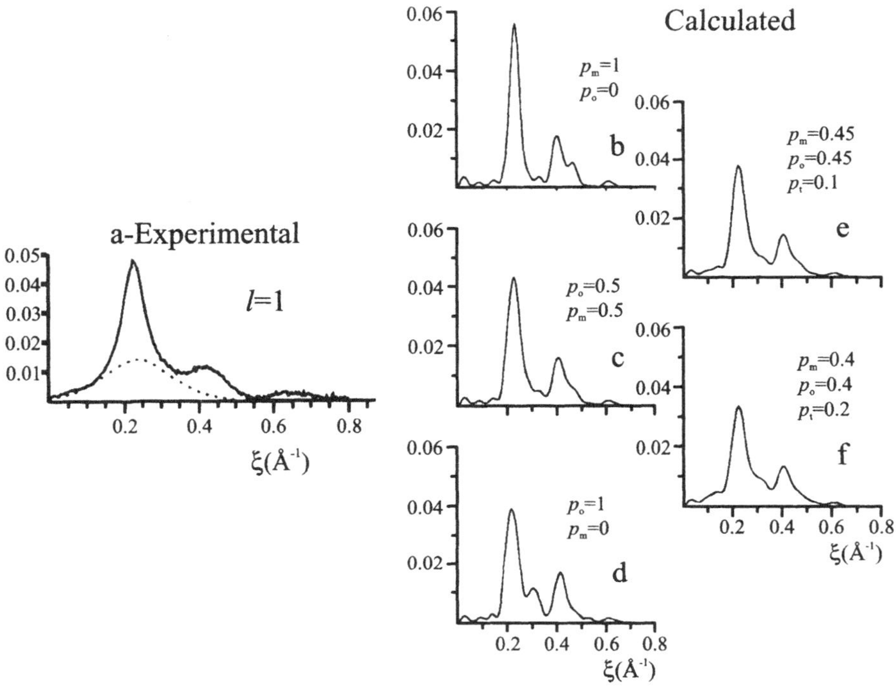

Fig. 16 (**a**) Experimental X-ray diffraction intensity distribution along the first layer line ($l = 1$) of a stretched sample of EP copolymer in the pseudo-hexagonal mesomorphic form (*solid line*). The *dashed line* indicates the contribution of amorphous phase. and 18° are also present, indicating the presence of (**b**)–(**d**) X-ray diffraction intensity distribution along the first layer line calculated for small aggregates of EP copolymer chains, where consecutive b_0c_0 layer of 4–6 chains of kind shown in Fig. 15, are faced along the a_0 lattice according to orthorhombic-like (Fig. 15A), monoclinic-like (Fig. 15B,B') and triclinic-like (Fig. 15C,C') models, with probabilities p_o, p_m and p_t, respectively. (Reprinted with permission from [195]. Copyright 1996 by the American Chemical Society)

culated patterns, and will be referred in the following as orthorhombic-like, monoclinic-like and triclinic-like ideal models.

A comparison between the experimental diffraction profile along the first layer line and those calculated for disordered models of structure in [195] is reported in Fig. 16. The X-ray diffraction intensity is calculated for disordered models of structure, characterized by small aggregates of EP copolymer chains, where consecutive b_0c_0 layers of chains of the kind of Fig. 15, are stacked along a_0 according to orthorhombic-like (Fig. 15A), monoclinic-like (Fig. 15B,B') and triclinic-like (Fig. 15C,C') ideal models, with probabilities p_0, p_m and p_t, respectively.

In all cases, a good agreement with experimental intensity distribution data is apparent, indicating that short-range correlations between the chains

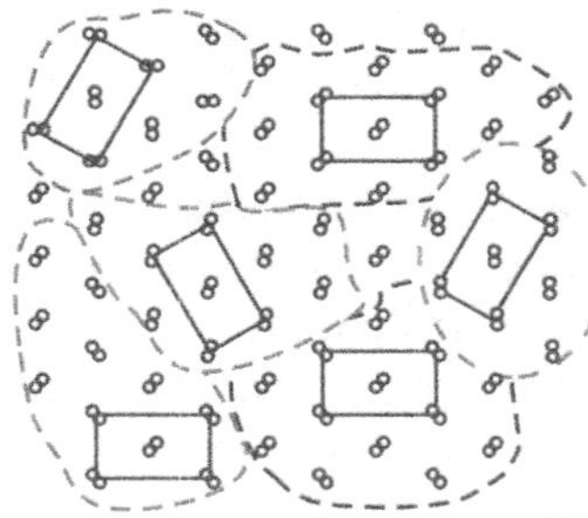

Fig. 17 The mosaic-like structure of crystalline micro-aggregate of EP copolymer chains in the pseudo-hexagonal form, where "ordered" domains of chains locally arranged as in the orthorhombic-, monoclinic- and triclinic-like model structures of Fig. 15, are assembled together. Long-range positional order of chain axes placed at nodes of a pseudo-hexagonal lattice is maintained. Regions enclosed in a different loop delineate domains of chains ordered in the short-range according to the structural models of Fig. 15

similar to those present in the orthorhombic, monoclinic and triclinic models are present in the small aggregates of mesomorphic form. These correlations rapidly fade away as the interchain distances increase. In other terms, the mesomorphic form of EP copolymers may be described as aggregates of clusters chains arranged as in the orthorhombic- monoclinic- or even triclinic-like structures, to form larger bundles, in a mosaic-like structure, as schematically shown in Fig. 17; the interference between different clusters in the mesomorphic aggregate would produce only a background on non-equatorial layer lines [195].

According to the analysis of [195], the mesomorphic hexagonal form of EP copolymers is characterized by the following structural features:

1. Long-range positional order of chain axes, placed at the nodes of a pseudo-hexagonal lattice, accounts for position and intensity of the equatorial reflections.
2. The presence of methyl groups in the small crystalline aggregates of EP copolymers necessarily introduces some conformational disorder. The conformational disorder consists in small deviations of the backbone dihedral angles close to the pendant methyl groups from 180°; these deviations from the *trans* state alleviate intramolecular and intermolecular strains and are easily digested in the pseudo-hexagonal lattice of EP copolymers. The chain portions in the crystalline aggregates, although remaining extended, with a mean periodicity close to 2.54 Å, would be rather wavy as a consequence of the presence of conformational disorder. Conformational disorder accounts for the experimental ratio between the integrated intensities of the main peaks on the first layer line and on the equator.

3. A local structure similar to ideal orthorhombic- and monoclinic-like structures shown in Fig. 15A and Fig. 15B,B' is retained at short-range distances, up to few tens of angstroms (2–3 times the lattice parameters). In other terms, on a local scale, the chains are arranged according to the orthorhombic-, monoclinic- or triclinic-like models of Fig. 15.
4. Translational disorder of chains parallel to c axes is present, as indicated by the broadness of non-equatorial peaks along the ξ direction and the fact that in Fig. 14B the halo on the second layer line is broader than the halo along the first layer line. This disorder was modelled by introducing in the Fourier transform calculations a paracrystalline parameter playing the role of decreasing the interference term along ζ between couples of atoms while increasing their distance in the plane normal to chain axes [195].
5. Rotational disorder of the chains around the chain axes is present as indicated by the presence of diffuse scattering subtending the Bragg reflections along the layer lines. This disorder was modelled introducing a paracrystalline parameter which affects the "ideal" angular position of the chains in the orthorhombic- monoclinic-or triclinic-like bundles. The loss of correlation in the relative angular position of couples of chains with respect to their "ideal" relative arrangement, was assumed to increase with increasing their distance [195]. The comparison of experimental and calculated X-ray diffraction intensity of bundles of EP copolymer chains indicated that the range of angular displacement disorder of chains about their axis is very small.

In conclusion, the pseudo-hexagonal form of EP copolymers corresponds to a solid mesophase of class ii). The small-sized crystals include large amounts of structural disorder and play an important role in elasticity of EP copolymers. These small crystals form upon stretching and melt when the tension is released, acting as physical knots of the elastomeric network. Since they are highly interconnected via the entangled amorphous chains (tie chains), the formation of crystalline knots prevents viscous flow during application of tensile stress and ensures the recovery of the initial dimensions of the sample upon releasing the tension [198].

5.5 Alternating Ethylene-Norbornene Copolymers

The copolymers of ethylene with norbornene produced with metallocene catalysts [199] are interesting thermoplastic materials characterized by high glass transition temperatures, good chemical and thermal resistance and excellent transparency [200, 201]. The properties of these copolymers depend on the comonomer composition, the distribution of the comonomers and the chain stereoregularity.

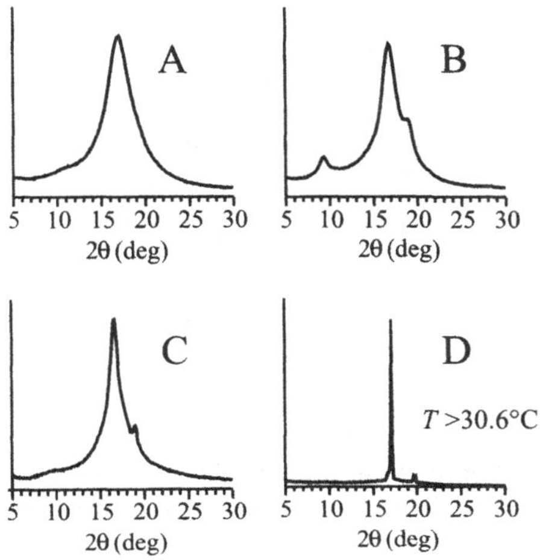

Fig. 18 X-ray powder diffraction profiles of ethylene-norbornene copolymer samples with norbornene content close to 50% (A)–(C), and solid norbornane in f.c.c. polymorph stable a $T > 30.6\,°C$, (D) redrawn from [121]. Sample (A) is a random copolymer and amorphous. Samples (B) and (C) have a prevailing alternating constitution and are both crystalline. Sample (C) is essentially atactic, whereas sample (B) has a prevailing di-isotactic configuration. (Reprinted with permission from [121]. Copyright 2003 by the American Chemical Society)

Recently, it has been shown that alternating ethylene-norbornene (EN) copolymers are crystalline [202–206] and that this crystallinity is not necessarily related to a regular succession of configurations of stereoisomeric centers in the norbornene units [121, 207].

The X-ray diffraction patterns of EN copolymers [121] is dominated by the presence of a broad halo centered at $2\theta \approx 16.7°$, which may be more or less sharp, depending on the sample, as shown in Fig. 18A-C. In particular the peak at $2\theta \approx 16.7$ is rather narrow for the EN copolymers of Fig. 18B and 18C. In the latter diffraction patterns narrow peaks of small intensity at $2\theta = 9.3$ and 18° are also apparent, indicating the presence of small crystals of dimensions less than 55 Å.

In the structural analysis of the crystallinity of EN copolymers of [121] it was recognized that the X-ray diffraction profiles of EN copolymers are all reminiscent of the diffraction profiles of plastic crystals like, for instance, that of norbornane (at temperatures $> 30.6\,°C$) [208] and adamantane [209], characterized by the packing of spherical molecules arranged in a face-centered cubic (f.c.c.) lattice. The X-ray diffraction profile of norbornane at $T > 30.6\,°C$ is shown in Fig. 18D, as an example. For the f.c.c. polymorph of norbornane, in fact, the powder diffraction pattern is dominated by an intense and narrow peak at $2\theta \approx 17°$, corresponding to (111) crystallographic planes and to an average distance between the spherical motifs of ≈ 6.17 Å [208].

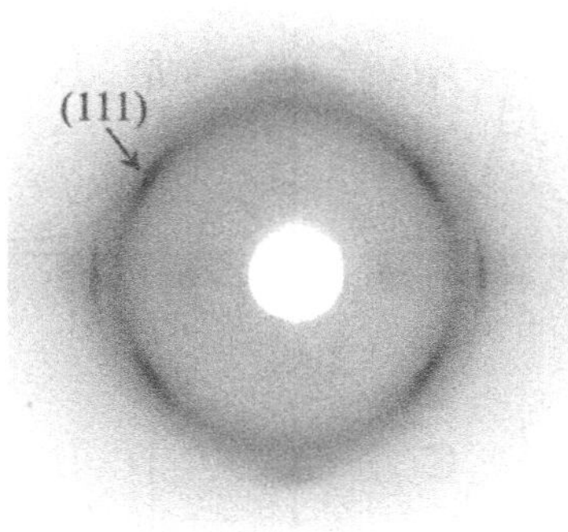

Fig. 19 X-ray fiber diffraction pattern of a stretched EN alternating copolymer sample with atactic configuration. The (111) reflection at $2\theta = 16.7°$ is indicated

This consideration led to hypothesize that the microcrystals of the alternating EN copolymers could be characterized by a similar face-centered mode of packing with ball-like norbornene units arranged in a face-centered paracrystalline lattice, at average distance ≈ 6.57 Å. The polymer chains in the paracrystalline aggregates could be organized in fringed micelle bundles [121].

The analysis of X-ray fiber diffraction pattern of an oriented fiber of an alternating EN copolymer sample (shown in Fig. 19) indicated that the main peak in the powder patterns at $2\theta = 16.7°$ is a first layer line reflection and that the identity period along the fiber axis corresponds to $c = 8.9$ Å. From the reflections observed in the X-ray powder and fiber diffraction patterns of alternating EN copolymer, a nearly tetragonal unit cell with a and b axes close to 9.4 Å and c close to 8.9 Å has been proposed [121, 207].

EN copolymers chains with norbornene content close to 50% may contain various types of constitutional and configurational units [210]. In the portions of chain with an alternating constitution, the configuration of the two chiral backbone atoms in each E-N-E unit is always SR or RS (Fig. 20), in agreement with a constant *cis-exo* insertion of norbornene molecules [210]. The corresponding N-E-N diads may be *meso* or *racemic* (Fig. 20). If all the diads along the chain were *meso*, the alternating copolymer would be diisotactic, whereas if all the diads were *racemic*, the alternating copolymer would be disyndiotactic [67].

Geometrical and conformational energy calculations have shown that nearly extended conformations are energetically feasible both for diisotactic and disyndiotactic EN chains [121, 207]; they correspond to two-fold helical s(2/1)m and glide plane tcm symmetries for the diisotactic and disyndiotactic configurations, respectively [67]. Moreover, both conformations account for the experimental chain axis period $c = 8.9$ Å and present distances be-

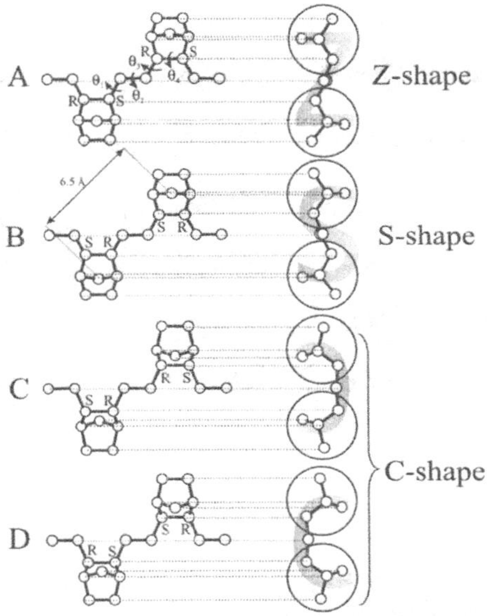

Fig. 20 (A,B) *Meso-* and (C,D) *racemo-* norbornene-ethylene-norbornene (N-E-N) sequences, in the lowest energy conformation. R and S indicate the configuration of the chiral carbon atoms of the norbornene units. Depending on the succession of the configurations R and S the *meso* diads may assume Z- or S-shapes (A,B), whereas the *racemic* diads may assume a C shape with different orientation (C,D). (Reprinted with permission from [121]. Copyright 2003 by the American Chemical Society)

tween the centers of mass of consecutive norbornene units close to 6.5 Å (Fig. 20) [121, 207].

Various kinds of limit-ordered and limit-disordered model structures have been proposed, arising from the packing of diisotactic and disyndiotactic chains, while maintaining long-range order in the average positioning of the norbornene rings, placed at the nodes of a face-centered lattice [207]. A limit-disordered model containing different kinds of statistical disorder is shown in Fig. 21A,B as an example [121]. It may be described, on average, by a tetragonal unit cell with axes $a = b = 9.4$ Å and $c = 8.9$ Å and the space group $I4/mmm$ (Fig. 21B).

In the structural model of Fig. 21A,B in each site of the lattice isotactic chains with Z- and S-shapes are present with the same probability. Moreover, chains rotated by 90° around the chain axis are present with the same probability. Differently oriented chains leave norbornene units basically in the same positions. This disordered model is characterized by positional order of the barycenters of norbornene units and orientational disorder, due to the fact that the chain may connect with equal probability a given norbornene unit with any of its neighbors. The X-ray diffraction profile calculated for this

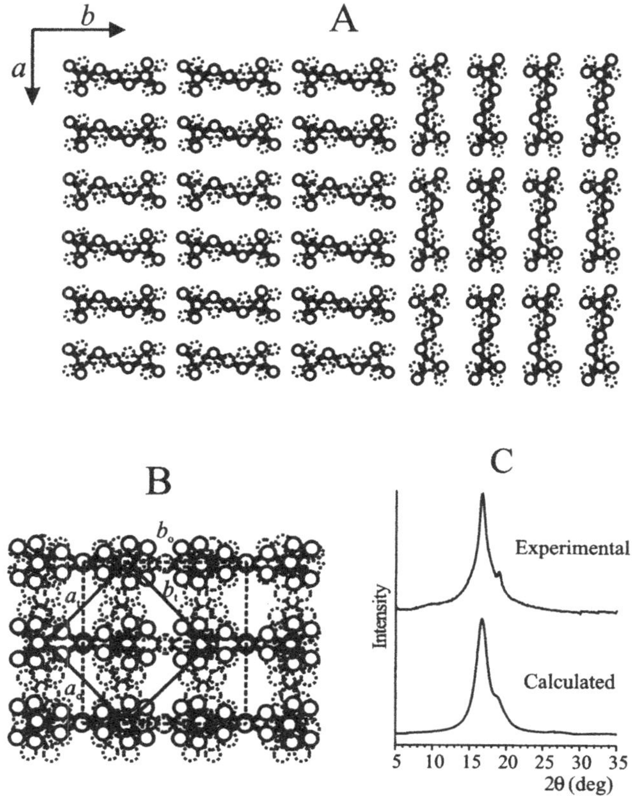

Fig. 21 (**A**) Disordered model of packing of alternating isotactic EN copolymer chains characterized by the presence of domains containing *bc* layers of chains piled along *a*, stacked along *b* with *ac* layers piled along *b*. The chains in the *ac* layers are rotated around the chain axes by 90° with respect to the chains in the *bc* layers. Each domain (*bc* layers piled along *a* and *ac* layers piled along *b*) present the statistical S-Z shape disorder, i.e. in each site of the lattice, chains with Z-shape (*continuous lines*) or S-shape (*dashed lines*) can be present with the same probability. (**B**) Limit-disordered model characterized by the statistical presence of *bc* layers and *ac* layers of chains perpendicular each other in the tetragonal unit cell, according to the statistical space group $I4/mmm$. The statistical S-Z shape disorder and statistical shifts of *bc* layers along *b* by 0 or $b/2$, and of *ac* layers along *a* by 0 or $a/2$ are also present. (**C**) Comparison between calculated X-ray diffraction profile for the limit-disordered model B (space group $I4/mmm$) and the experimental X-ray powder diffraction profile of alternating EN copolymer sample of Fig. 18C [121]

model is in a good agreement the experimental powder diffraction profile of an alternating EN copolymer sample, as shown in Fig. 21C [121].

Whichever the configuration of the diads, all the alternating copolymer macromolecules have a similar shape (Fig. 22). It has been shown that partial three-dimensional order may be obtained in the EN copolymers, guided by the positioning of the ball-like norbornene units in a f.c.c. "paracrystalline"

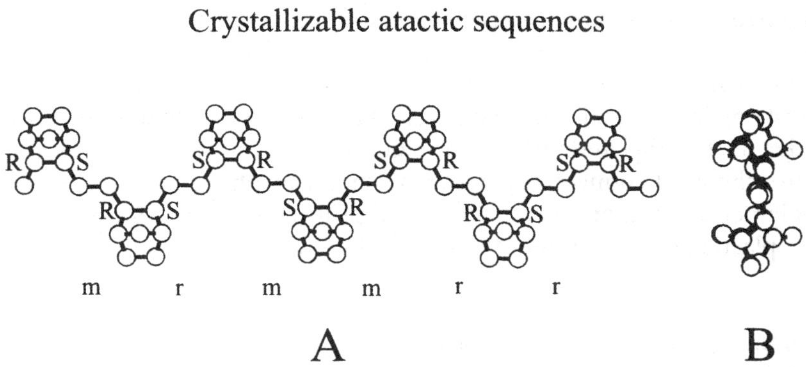

Fig. 22 (**A**) Projection parallel and (**B**) perpendicular to chain axis of EN alternated copolymer chain with atactic configuration. *meso* (*m*) and *racemic* (*r*) N-E-N diads follow each other at random in extended, low-energy conformation

lattice, even though the polymers are not stereoregular, provided that they have a regular alternation of the comonomeric units [121, 207].

The ordered positions of the norbornene units in the crystalline domains is shown in Fig. 23. It is apparent that the ball-like units are organized on a face centered-lattice. In this projection, only one of the possible ordered positions of the copolymer chain is shown. As also shown in Fig. 21B, the ethylene units, which connect the norbornene units, may assume different positions along the *a* and *b* axes, producing orientational disorder of the chains and of the quasi-spherical norbornene units. However, the barycenter of the norbornene units remain organized on a face-centered lattice, producing the strong 111 reflection [121].

In conclusion, the crystalline phase of alternating EN copolymers corresponds to a solid mesophase of kind ii) where long-range periodicity is

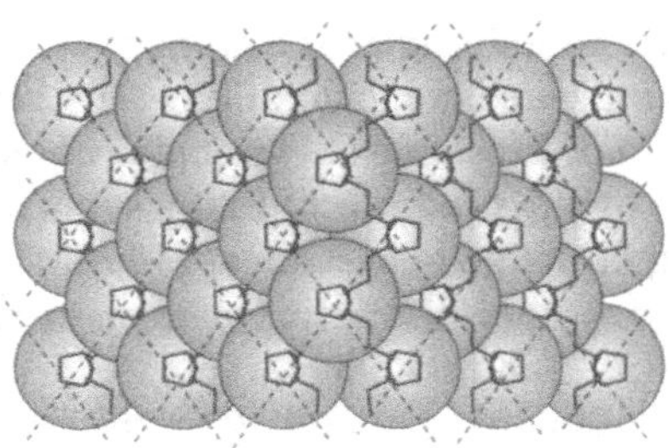

Fig. 23 Projection along the *c* axis of the structure of alternating EN copolymer, showing the ordered packing of ball-like norbornene units in a face-centered lattice. Only one of the possible ordered positions of the copolymer chain is shown. The *dashed lines* are the trace of strongly diffracting (111) planes of the face-centered lattice. (Reprinted with permission from [121]. Copyright 2003 by the American Chemical Society)

maintained in three dimensions for a not-point-centered structural motif, the center of mass of norbornene units, but not for the atomic positions. The similarity between the alternating EN copolymer and the plastic crystals, rests mainly related to the structure due to the presence of orientational disorder of the ethylene units and the quasi-spherical norbornene units. In fact, the chemical connectivity between the quasi-spherical motifs, through the polymeric chain, produces mechanical properties typical of conventional thermoplastic polymers rather than of plastic crystals.

5.6 Pseudo-Hexagonal Form of Polyethylene at High Pressure and Temperature

At ambient conditions, polyethylene (PE) crystallizes in the orthorhombic form with chains in *trans*-planar conformation [194]. At elevated pressures (of the order of hundreds of MPa) and high temperatures, polyethylene crystallizes in extended-chain crystals with lamellae which can be several microns thick, well distinguishable from the chain-folded lamellae obtained at atmospheric pressure [211–214]. In fact, in the normal crystallization conditions, PE lamellae are characterized by thickness of the order of 100–200 Å, which may extend at the utmost by a factor of 3–4 times the initial value through subsequent annealing treatments [215, 216]. The formation of extended-chain type lamellae has been related to the appearance of an intermediate phase, thermodynamically stable above 3.6 kbar and 200 °C [15–17, 123, 217, 218]. It has been suggested that in this phase, chains achieve a high mobility, allowing the crystals to grow simultaneously in the lateral and thickness direction, leading to formation of extended chain crystals. PE samples crystallized in extended chain lamellae show crystallinity close to 100%, melting temperature close to thermodynamic melting temperature and mechanical properties more similar to those of common polycrystalline materials of low molecular weight, than to those of thermoplastic polymers [211–214].

A hexagonal structure containing a high amount of structural disorder has been proposed for this intermediate form of PE [219–221]. The X-ray diffraction patterns of the form of PE stable at elevated pressures and temperatures is indeed characterized by the presence of Bragg spots along the equator and diffuse scattering along ill-defined layer lines [219–221].

In [221], a typical X-ray fiber diffraction pattern of an uniaxially oriented PE sample, taken at 285 °C and 9 kbar is reported. Three equatorial Bragg reflections corresponding to 100, 110 and 200 reflections of a hexagonal lattice with $a = 4.81$ Å and some off-equatorial diffuse scattering at $\zeta = 0.42$ and 0.80 Å^{-1}, corresponding to a first and a second layer line, respectively, are present. The layer line diffuse scattering indicates an average chain periodicity close to 2.4 Å.

Raman spectroscopy indicates that the chain conformation of PE chains in the hexagonal polymorph is not much different from that in the amorphous state [222]. These observations, along with the measured low melting entropy [223–226], indicate the presence of a high amount of structural disorder in the hexagonal form of PE. Owing to the high amount of structural disorder and the high mobility of PE chains, the term "liquid crystalline" has been often used to address the hexagonal form of PE stable at high pressures and temperatures [227].

The transformation of the orthorhombic form into the hexagonal form at temperature above 200 °C and elevated pressure produces a chain contraction of 6–8% [217, 224]. Raman spectra of PE in the hexagonal polymorph stable at high pressures indicate the presence of a high amount of C – C skeletal bonds in the gauche conformation [222]. Thus, although never directly observed, the presence of large amounts of kink conformational defects of the kind $TG^{\pm}TG^{\mp}$ within an otherwise fully *trans*-planar chain, was inferred [219]. The presence of kink defects preserves the straightness of the chain and shortens the chain, according to the experimental data. One kink produces, indeed, a shortening by one C – C repeat distance. The observed contraction of chain length would imply the presence of about 6–8 kinks per 100 C atoms [228]. The chain conformation, although highly disordered, would retain, overall, an elongated shape in the mesomorphic form of PE, sufficient to maintain the parallelism of chain axes within small bundles of close neighboring chains, arranged in a hexagonal lattice.

Indirect observations concerning the presence of kink defects within the hexagonal form of PE came from independent IR experiments performed at atmospheric pressure on γ irradiated PE samples [228] and on ultra drawn PE yarns kept under tension at temperatures higher than the normal melting temperature of PE at atmospheric pressure [229].

In fact, the hexagonal form of PE may be also obtained above a critical pressure, by heating γ irradiated PE samples above a certain temperature [230]. The orthorhombic to hexagonal transition temperature is a function of γ ray dose, and a sufficiently high irradiation dose makes the radiation induced hexagonal phase of PE stable even at atmospheric pressure [228, 230]. The radiation-induced hexagonal phase shows IR spectra with the characteristic bands of kink defects [228].

Moreover, it has been shown that at atmospheric pressure a phase transition of the orthorhombic form into a hexagonal form may occur by heating ultra-drawn PE samples a few degree above the normal melting temperature, if the sample ends are tightly bound, to prevent retraction [229]. This transition has been followed by in situ analysis through X-ray diffraction technique and IR and Raman spectroscopy. The X-ray fiber diffraction pattern of the so-obtained hexagonal phase presents several analogies with those obtained for uniaxially oriented PE samples at high pressure and temperature. In addition, direct evidence of the presence of *gauche* bonds, and in particular of kink defects, also in this conditions are provided [229].

In [221] the X-ray diffraction patterns of PE fibers in the high-pressure hexagonal form have been modeled assuming complete translational disorder along *z*, rotational disorder of chains around their axes and conformational disorder. The Fourier transforms of disordered structural models were calculated as a function of intra- and inter-molecular parameters related to the presence of conformational disorder and the relative arrangement of close neighboring chains (short-range correlation disorder). The results of the calculations were compared to experimental X-ray diffraction data.

The analysis reported in [221] allowed to address basic features which characterize inter- and intra-molecular disorder of PE in the hexagonal mesomorphic form. A long-range order in the position of chain axes arranged in a two-dimensional hexagonal lattice is present. In these large hexagonal bundles of parallel chains, local ordered domains, characterized by a structure similar to that of the usual orthorhombic form of PE are present, as shown in Fig. 24. The range of such inter-molecular correlation (τ, in Fig. 24) was estimated to be about 4–5 Å. The size of these ordered regions would be limited in the chain axis direction and in the lateral direction because of the presence of conformational disorder in the chain.

The presence of kink conformational defects in a prevailing *trans*-planar conformation accounts for the shortening of the chain axes from 2.5 Å of the full *trans*-planar conformation to ≈ 2.4 Å. Introducing considerations *a posteriori* concerning the average fraction of conformational defects and the

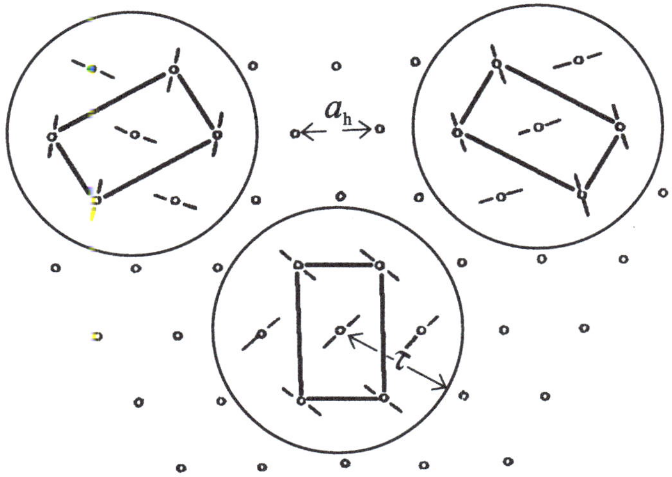

Fig. 24 Model of the local intermolecular structure in the high-pressure phase of PE. A local structure similar to that of the usual orthorhombic form is retained within the small range $\tau(\approx 4–5$ Å). (Adapted from [221])

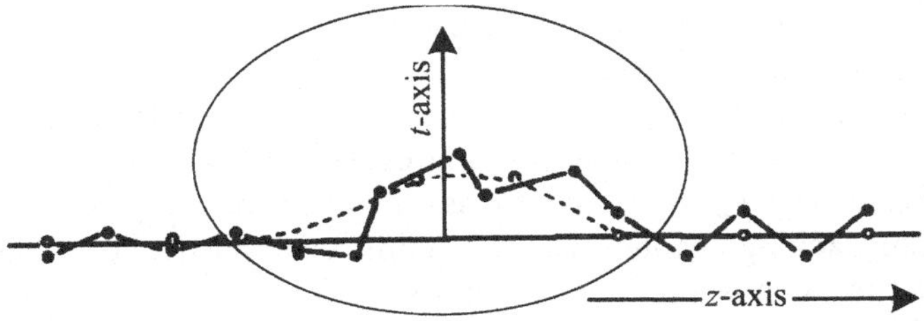

Fig. 25 Schematic drawing of a PE chain with a kink defect shortening the average chain periodicity from 2.55 Å (corresponding to the all-*trans*conformation) to values close to 2.4 Å. The defective region of the chain is *encircled*. The t parameter expresses the lateral encumbrance of kink defect along an axis perpendicular to z (chain axis). The chain portions of PE aside the defect are co-axial. (Adapted from [221])

standard deviation of the lateral positions of defective units, the lateral encumbrance, t, of kink defects (defined in Fig. 25) was esteemed to be about 1.6 Å, in a good agreement with the lateral encumbrance of the kink model proposed in [105] by Petraccone, Allegra and Corradini.

The hexagonal form of PE stable at high pressure, may be classified as a solid mesophase of class ii) characterized by long-range order in two dimensions of the chain axes placed at the nodes of a hexagonal lattice and no periodic order along the third dimension, because of occurrence of conformational disorder.

5.7 Poly(tetrafluoroethylene)

Poly(tetrafluoroethylene) (PTFE) at atmospheric pressure shows a peculiar polymorphic behaviour involving three crystalline phases as the temperature increases [50]. The three polymorphs are usually denoted as form II, stable at temperatures lower than 19 °C, form IV stable at temperature between 19 and 30 °C and form I stable at temperatures higher than 30 °C [50]. At 19 °C form II transforms into form IV which in turn transforms into form I at 30 °C, resulting in a step-like increase of disorder in the crystals [49].

X-ray fiber diffraction patterns of form II ($T < 19$ °C) present narrow Bragg reflections along well-defined layer lines, indicating long-range, three-dimensional order. According to Bunn and Howell [10], the X-ray diffraction intensity distribution in the low temperature form may be interpreted in terms of the close packing of helical chains containing 13 CF_2 units in 6 turns. Successively [49, 231, 232], Clark et al. have found that the assumed commensurable 13/6 helix is a non-commensurable helix.

Chains in nearly 13/6 conformations are packed in a nearly hexagonal lattice with $a' = b' = 5.59$ Å, $\gamma' = 119.3°$ [11]. According to [52], form II of PTFE is characterized by rows of isomorphous helices alternating with rows of chains of opposite chirality, in a nearly hexagonal arrangement (triclinic unit cell, 2 chains/unit cell, see also [51,233–235]).

Pierce *et al.* [49] have shown that the 19 °C transition from form II into form IV corresponds to slight untwisting of helical chains, from 13/6 helix into 15/7 helix (15 CF_2 units in 7 turns). At onset of the phase transition, the molecular packing changes from a triclinic ordered structure into a hexagonal structure with larger inter-chain distances, including large amount of structural disorder. The 15/7 helices have a 3/1 symmetry, consistent with the trigonal symmetry of the unit cell of form IV. According to Clark [232], the chain repetition of PTFE chains in form IV corresponds to $c = 19.5$ Å, and at 20 °C, the chain axes would be placed at the nodes of a hexagonal lattice with $a' = b' = 5.66$ Å. Also form IV is characterized by a packing of right- and left-handed helices. In particular, the hexagonal packing is characterized by rows of isomorphous chains stacked with rows of chains of opposite chirality [53, 236].

Bunn and Howells [10] have observed that when the temperature is raised above 19 °C, the X-ray diffraction patterns of PTFE fibers show that only a few Bragg reflections on the equator and on the 7th, 8th and 15th layer lines remain and that large amount of diffuse scattering, localized along well-defined layer lines, are present (see also [232]). The presence of the diffuse scattering in the X-ray diffraction patterns of form IV was attributed to the onset of disorder in the structure involving rotation of chains around chain axes and translation of chains along their axes. Klug and Franklin [155] argued that such kind of diffuse patterns may arise from random screw displacements implying a combination of translation and rotational motion of chains about their axes along defined helical directions.

At 30 °C form IV transforms into form I which is characterized by the presence of larger amounts of disorder, the chain conformation becomes more irregular and long-range order is maintained only in the parallelism of chain axes and in their pseudo-hexagonal arrangement [11].

In [11] and [232]a, accurate X-ray diffraction measurements on PTFE fibers at different temperatures, revealed that form IV is characterized by three kinds of layer lines, those with sharp spots only (equator and 15-th layer lines), those with sharp spots and diffuse halos (6,7 and 8-th layer line), and those with diffuse halos only (all remaining layer lines); at 30 °C transition the sharp spots disappear except for those on the equator and on the 15th layer line. X-ray diffraction patterns of form I present, indeed, sharp reflections only along the equator and the 15-th layer line and diffuse halos in an intermediate region of the reciprocal space placed between the 7-th and 8-th layer lines [11, 237]. At temperatures higher than 150 °C, also the sharp peaks on the 15th layer line disappear [54].

Clark and Muus extended the diffraction theory of helical structures by Cochran, Crick and Vand to include the effect of specific types of disorder in crystals of helical polymers and applied the theory to the analysis of X-ray diffraction patterns of PTFE in forms IV and I [11]. Translational and rotational disorder and screw displacements of chains about their axes have been considered.

It was argued that the X-ray diffraction intensity distribution of form IV may be accounted for by disorder mainly consisting in small angular displacements of the chains around their axes; the angular displacements would become much larger in the form I at temperatures higher than 30 °C transition [11]. The occurrence of random screw displacements in form IV as proposed by Klug and Franklin [155] was ruled out by this analysis, because this kind of disorder does not account for the experimental X-ray diffraction intensity distribution on the 6-th, 7-th and 8-th layer lines, characterized by the simultaneous presence of Bragg and diffuse scattering and for the absence of any continuum scattering along the 15-th layer line. The mechanism of crystal disordering at 19 °C was then related to the high mobility of chains in PTFE crystals found in preceding studies by Hyndman and Origlio [238] using solid-state NMR techniques. A possible mechanism for crystal disordering involving small twisting and untwisting of the helices around a mean value, coupled with a gradual untwisting of the helical molecules was proposed in [11].

The amount of disorder increases in form I. In particular the presence of conformational disorder, due to activation of helix reversals at temperatures higher than 30 °C has been proposed [12, 239–246]. The effect of helix reversals on the X-ray diffraction intensity distribution in PTFE was firstly explicitly considered by Corradini and Guerra in [12].

Helix reversals in [12] have been modeled maintaining consecutive helical stems of opposite chirality coaxial, the transverse radius of the resulting chain constant and the z coordinate (parallel to chain axis) of CF_2 groups in register as in the defect-free helices. It was shown that within these rigid constraints it is possible to model chains of PTFE containing helix reversals at a low cost of conformational energy, with bond lengths and valence angles around the reversals only slightly displaced from those in the prefect helix [12]. A model chain of PTFE containing a helix reversal is shown in Fig. 26.

In [12] and in later studies [245, 246] it was shown that the occurrence of helix reversal in PTFE 15/7 helical chains may account for some experimental observations, concerning the X-ray diffraction intensity distribution on the 7th and 8th layer lines. In particular, in [246] it was demonstrated that the number of helix reversals in form I increases with increasing temperature. In fact, the change observed in the X-ray diffraction pattern with increasing temperature can be accounted for by the increase of the concentration of helix reversals [246].

The experimental X-ray diffraction profiles of an uniaxially oriented PTFE sample are reported in Fig. 27 at different temperatures, as a function of the reciprocal coordinate ζ, redrawn from [246]. The value of ξ reciprocal coordinate is fixed close to the position of intensity maximum on the 7th and

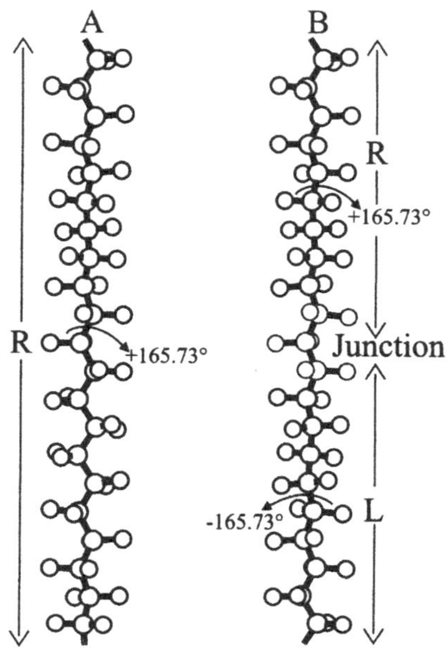

Fig. 26 (**A**) PTFE chain in 15/7 helical conformation. (**B**) PTFE helical chain incorporating a single helix reversal

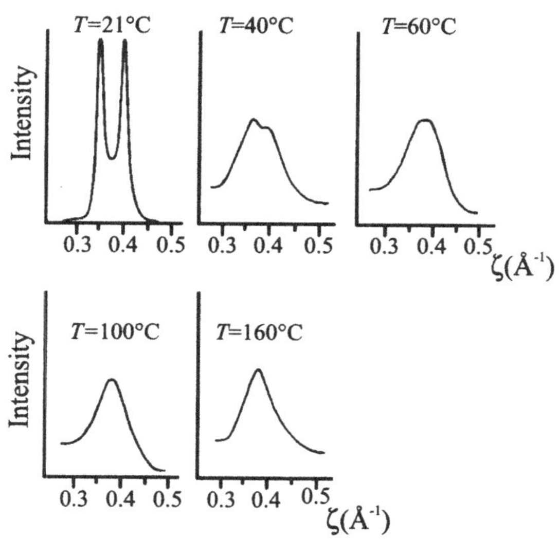

Fig. 27 Experimental X-ray fiber diffraction patterns of PTFE at $\xi = 0.20$ Å^{-1} in the region of 7-th and 8-th layer line as a function of the reciprocal coordinate ζ. (Reprinted with permission from [246]. Copyright 1988 by the American Chemical Society)

8th layer lines. It is apparent that, while below 30 °C the 7th and 8th layer lines are well defined, with increasing temperature the two layer lines become ill defined and approach each other, merging into a single broad-layer line centered in between; with further increase of temperature (up to 160 °C), the profile of the single-layer line becomes increasingly narrow. This behavior could be explained as due to a thermally activated process leading to an increase of concentration of helix reversals as the temperature increases.

In Fig. 28 the calculated X-ray diffraction profiles for model chains containing different concentrations of helix reversals redrawn from [246], are shown; it is apparent that the presence of 1 reversal every 20 and 10 CF_2 units could account for the experimental patterns of form I at low (30–40 °C) and high (160 °C) temperatures, respectively (Fig. 27).

Corradini, De Rosa *et al.* [245] have evaluated in explicit manner the effect of various kinds of disorder on the X-ray diffraction intensity distribution, performing Fourier transform calculations on model bundles of chains of finite size containing specific kind of disorder. Disorder due to the presence of random translational displacements of chains along the chain axes, random rotational displacements around their axes with respect to an average position (already treated by Clark and Muus [11]), random placement of left- and right-handed helices in the positions of the pseudo-hexagonal lat-

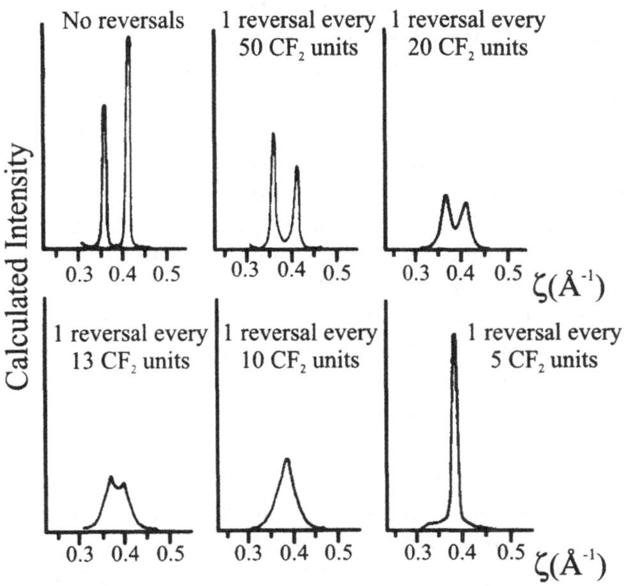

Fig. 28 Calculated X-ray fiber diffraction profiles at $\xi = 0.20$ Å^{-1} as a function of ζ for chain models of PTFE containing various concentration of helix reversals. (Reprinted with permission from [246]. Copyright 1988 by the American Chemical Society)

tice and the presence of helix reversal defects, have been analyzed [245]. The results of [245] have indicated that, in agreement with arguments of Clark and Muus [11], the disorder in form IV mainly consists in the occurrence of small angular displacements of chains around their axes and of a little, if any, translational disorder of chains parallel to their axes. The chains are organized, at least in the short range, in rows of isomorphous helices, packed with rows of chains with opposite chirality. In form I, besides the disorder of form IV, helix reversals defects are present. In this form, at high temperatures, any residual translational order parallel to chain axes and orientational order around chain axes, would be lost [245].

Later on, Kimming, Strobl and Stün [247] performed a detailed analysis of static and dynamic scattering measurements and of infrared and Raman spectra on PTFE oriented samples at different temperatures. In addition, in [247], the influence of different kinds of disorder on the X-ray diffraction intensity distribution was also investigated. The analysis of diffuse scattering on the 7th and 8th layer lines essentially confirmed that an orientational short-range order is retained for sequences in adjacent chains in the intermediate form IV and that the degree of order possibly decreases with increasing the temperature in form I.

The degree of orientational order in form IV and form I is strictly related to the amount of helix reversal defects. Kimming et al. [247] have proposed that the helix reversals occur forming pairs, rather then isolated defects. The important role of twin helix-reversal defects for molecular dynamics in the form IV and form I of PTFE was pointed out. In contrast to single helix-reversal defects, indeed, these twin defects constitute a local perturbation which can move along the chain without affecting remote units. In other terms, occurrence of twin helix reversal defects would not affect the packing of remote units, being a lattice perturbation associated with a small space region; according to [247] a particular importance is assumed by twin helix reversals comprising 15 CF_2 units, which possess a particularly low formation energy, since they retain the long-range twist along the chains. It was shown that the concentration of twin defects increases steeply with temperature in the intermediate form, between 19 and 30 °C, and levels off in the high-temperature form as the temperature increases. It was argued that this behavior indicates a transition between a disordered state controlled by intra- and inter-molecular forces into one which is only determined by intramolecular potentials [247].

In addition, dynamic scattering measurements have indicated high mobility of PTFE chain segments in the solid state, which results from the formation and motion of helix reversal defects [247]. The reorientation of CF_2 groups occurs with rates in the range of 10–100 GHz, in the high-temperature forms, the presence of short-range orientational order in form IV leading to a slowing down of this motion [247].

In conclusion, long-range order of chain axes placed at the nodes of a two-dimensional lattice is present in all crystalline forms of PTFE up to high temperatures, the lateral packing of the helices being pseudo-hexagonal in Form II, hexagonal in form IV and form I. Transition of form II into form IV results from a slight untwisting of nearly 13/6 helices into 15/7 helices. Form IV is characterized by a small amount of disorder, mainly small rotational displacements of chains around their axes. As the temperature increases, also disorder increases. At 30 °C a first order phase transition occurs, leading to a highly disordered crystalline form, characterized by a high concentration of helix reversals, large rotational displacements of chains around their axes and ultimately, at high temperatures, by large longitudinal translation disorder of chains along the chain axes. As a consequence of the presence of a large amount of helix reversal defects, long-range order of atomic positions parallel to chain axes is lost in form I.

Therefore, the high temperature form of PTFE corresponds to a solid mesophase belonging to the class ii), characterized by long-range order of not-point-centered structural features, i.e. the chain axes which remain arranged parallel each other at nodes of a two-dimensional hexagonal lattice, up to high temperatures, whereas no order is present along the chain axis because of the conformational disorder. The high degree of disorder in form I, in particular the conformational disorder present at high temperatures, accounts for the very high melting temperature of PTFE, due, indeed, to the low melting entropy of PTFE mesomorphic crystals [248].

As a final remark, we point out an unusual property of PTFE solid mesophase at high temperatures, indicative of a plastically crystalline nature, due to the high mobility of chain axes in the crystalline state. As reported by Wittman and Smith [249], if a bar of solid PTFE is moved under weak pressure against a hot surface, for example a glass slide, an ultrathin layer becomes deposited. The crystallites in this thin film show a perfect uniform orientation, the chains being directed along the rubbing direction.

5.8
Alternating Ethylene-Tetrafluoroethylene Copolymers

Alternating ethylene-tetrafluoroethylene copolymer (ETFE) having a 50/50 comonomer composition crystallizes in two polymorphic forms. An orthorhombic form, stable at room temperature, and a disordered hexagonal form, stable at high temperatures. The crystal structures of both forms have been studied and the structural disorder present in both forms have been analyzed as a function of the composition [250–256].

Both crystalline forms of ETFE copolymers may be considered as mesomorphic forms [250–253]. In the low-temperature form, the chains are in the ordered *trans* planar conformation and are packed in a polyethylene-like orthorhombic lattice [250, 251]. However, intermolecular order is present only in the *ab* projection of the lattice, any intermolecular correlations in the relative positions of the atoms being absent due to the presence of nearly random relative displacements of neighboring chains along the *c* axis [253]. Fourier transform calculations of structure models have indicated that this disorder arises basically from the presence of constitutional disorder in the chains, due to the not perfect alternation of comonomer units [253, 256].

At high temperatures, this orthorhombic mesomorphic phase of ETFE transforms into a hexagonal mesophase characterized by chains still in the ordered *trans*-planar conformation packed in a hexagonal lattice [250–252]. Besides the disorder in the relative heights of the chains, disorder in the rotations of the chains around their axes develops at high temperature, inducing the hexagonal packing [250–252]. A long-range, three-dimensional order is maintained only along each *trans*-planar chain and in the hexagonal arrangement of the chain axes [252].

The transition from the orthorhombic form into the hexagonal form occurs at 100 °C for the perfectly alternating copolymer with 50/50 comonomer composition. The polymorphic transition is revealed from the X-ray diffraction patterns by the transformation of the two typical equatorial reflections of the orthorhombic form into a single reflection, which becomes narrower and narrower with increasing temperature [250–252]. It has been demonstrated that this transition is a topotactic, reversible transformation, generated by the trigger of disorder in the relative rotations of chains around their axes [252]. Small orthorhombic crystals having different orientations may transform into larger hexagonal crystals, which are characterized by a lower degree of local order (Fig. 29) because of the rotational disorder. Therefore, during this transition, larger, but more disordered crystals are formed with increasing temperature [252].

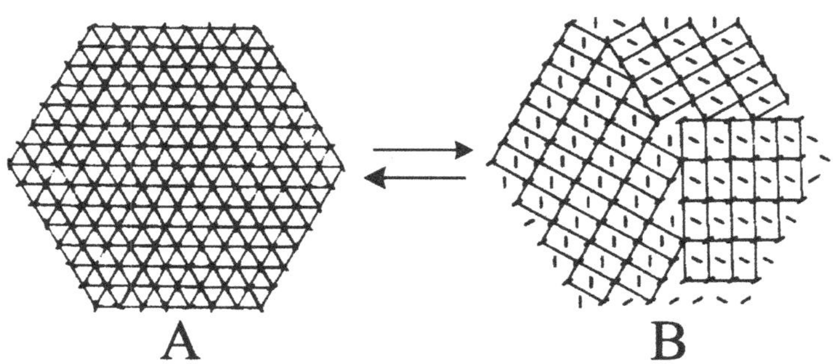

Fig. 29 Topotactic transition of alternating ethylene/tetrafluoroethylene copolymers from orthorhombic into hexagonal form. (**B**) Small orthorhombic crystals having different orientations transform into (**A**) larger hexagonal crystals, which are characterized by a lower degree of local order because of the rotational disorder. (Reprinted with permission from [252]. Copyright 1989 by Wiley)

In ETFE copolymer samples with compositions different from 50/50, a decrease of the transition temperature with increasing concentration of tetrafluoroethylene (TFE) comonomeric unit at values higher than 50%, has been observed [252, 255]. Only the hexagonal mesomorphic form has been, indeed, observed for ETFE copolymers with TFE contents higher than 80% [252]. Conformational energy calculations have shown that the increase of TFE concentrations, with respect to the 50/50 composition, induces an increase of intramolecular disorder, which produce local deviation from the trans-planar conformation [252]. The presence of these constitutional defects, with respect to the alternating ETFE copolymer with 50% composition, favors the development of the intermolecular rotation disorder, and, therefore, an increase of the stability at room temperature of the hexagonal mesomorphic form [252, 255]. Similar effects have been found in alternating 50/50 ETFE terpolymers, containing small concentrations (1–3 mol %) of a termonomer, as for instance, perfluoropropylperfluorovinylether, which also reduces the degree of crystallinity and the crystal sizes of the pseudo-hexagonal mesophase [255].

Similar hexagonal mesomorphic form has been described for alternating ethylene-chlorotrifluoroethylene (ECTFE) copolymers [254, 256, 257]. Also, in this case, chains in trans-planar conformation are packed in a disordered pseudo-hexagonal lattice with random disorder in the relative rotations of chains around their axes and in the relative displacements along the chain axis [254, 256].

6 Solid Mesophases with Long-Range Positional Order in Two or One Dimensions

6.1 Poly(ethylene terephthalate)

Poly(ethylene tephthalate) (PET) is a crystalline polymer characterized by a triclinic unit cell. The chain axis is 10.7 Å and corresponds to chains in fully extended conformation, with all dihedral bonds close to the *trans*-planar conformation (*ti* symmetry) [258]. As shown in Fig. 30A, in this conformation, the terephthalic residues are all coplanar.

Cold drawing of PET amorphous samples (at temperatures lower than the glass transition temperature, T_g) induce formation of a highly disordered metastable mesomorphic form, which transforms into the normal crystalline (triclinic) form by heating above T_g [18–20]. This form was early identified by Bonart [18], studying the stress induced crystallization of amorphous PET samples by X-ray diffraction techniques. The diffuse nature of X-ray diffraction patterns of samples obtained by stretching at room temperature

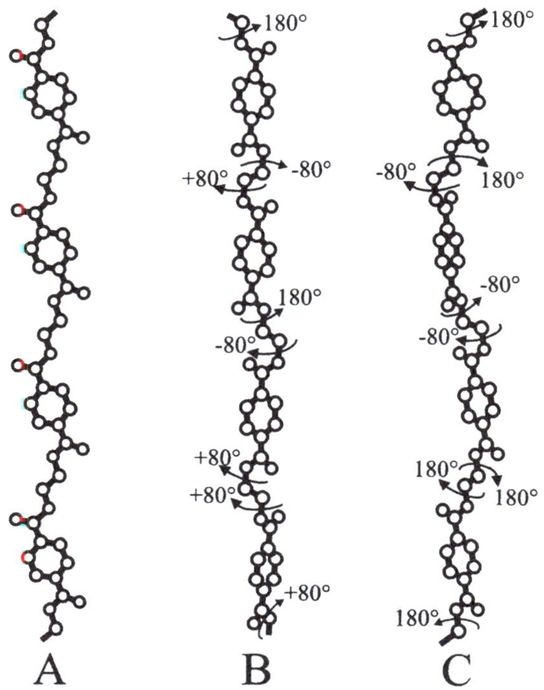

Fig. 30 (**A**) Chain conformation of PET in the triclinic form. Backbone dihedral angles are close to *trans* state and all phenyl groups are coplanar. (**B**)–(**C**) Disordered conformations of PET in the mesomorphic form, consisting in random sequences of monomeric units in different minimum energy conformations. Low minimum energy conformers are characterized by all backbone dihedral angles in the *trans* conformation, except for CO – O – C – C bonds which may assume in consecutive monomeric units values close to +80, – 80° or 180°

amorphous PET was explained by Bonart [18] in terms of a nematic hexagonal (columnar) structure followed by formation of a smectic structure at higher temperatures and, then, crystallization into the triclinic form at temperature higher than T_g. The term "paracrystalline" form was also used by Bonart [18–20] to address the disordered mesomorphic form of PET.

X-ray fiber diffraction patterns of PET in this solid mesophase present, indeed, Bragg reflections only on the meridian at values of $\zeta = 1/c$, $3/c$ and $5/c$ corresponding to a spacing of 10.3 Å (Fig. 31B); off the meridian only a broad equatorial halo is present (Fig. 31A).

Asano and Seto [259] interpreted the equatorial broad halo as due to overlap of four crystalline reflections of a paracrystalline monoclinic structure. The presence of the meridional reflections at $d = 10.3$ Å was explained by a tilt of 10° of c axis of the monoclinic unit cell with respect to the drawing direction; this is analogous to the triclinic structure of PET studied by Bunn [258], for which X-ray diffraction patterns of well-oriented fibers

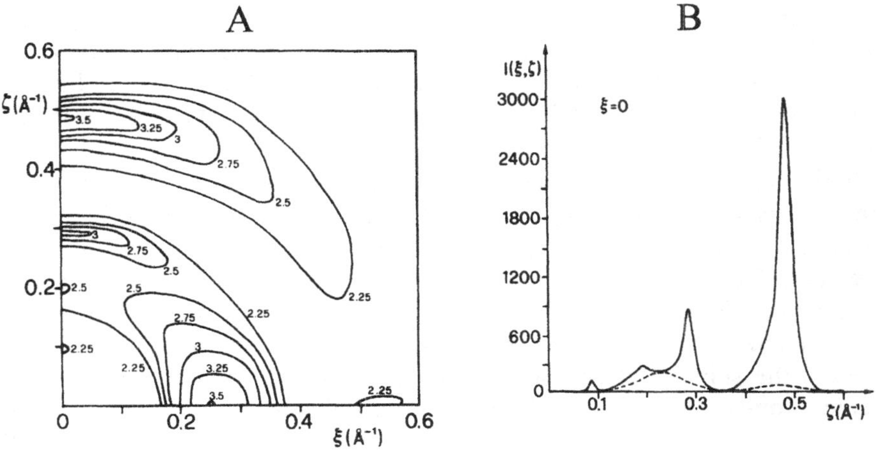

Fig. 31 (**A**) Experimental X-ray diffraction intensity $I(\xi,\zeta)$ of PET fiber in the mesomorphic form as a function of the reciprocal cylindrical coordinates ξ and ζ, and (**B**) corresponding profile along the meridian. In (**A**), contour lines with constant $\log[I(\xi,\zeta)]$ values are at regular intervals of 0.25 whereas in (**B**) the dashed line indicates the amorhous contribution. (Reprinted with permission from [260]. Copyright 1992 by the American Chemical Society)

present reflections displaced from the normal layer lines, indicating that the chain axes (corresponding to the *c*axis) of the triclinic crystals are tilted with respect to the fiber axis. Therefore, for the mesomorphic form of PET, the *c* axis (parallel to chain axes) of the monoclinic unit cell would be equal to $10.3/\cos(\tau)$, with τ the tilt angle, giving a chain periodicity $c = 10.5$ Å, only slightly shorter than chain periodicity of PET in the stable triclinic form [259].

According to Asano and Seto [259], the monoclinic structure of PET would be paracrystalline in the sense that the monoclinic symmetry would describe, on average (in the long range), a mosaic-like structure formed by the assembly of crystalline microdomains in the normal triclinic form of PET (in the short range) with different inclination.

A different model for the structure of PET in the mesomorphic form was proposed in [260], since the model given by Asano and Seto [259] does not account for the lack of diffraction (Bragg and diffuse scattering) aside the meridian and the equator.

The structural model proposed in [260] is based on accurate X-ray diffraction measurements over regions of the reciprocal lattice larger than those sampled in the preceding literature (shown in Fig. 31) and on geometrical and conformational analyses and calculations of Fourier transforms of model structures.

The X-ray diffraction patterns of Fig. 31A indicates that the not perfect alignment of chain axes parallel to fiber axis affects the shape of diffraction peaks, smearing the diffracted intensity in the reciprocal space over large arcs with constant values of $s = 2\sin\theta/\lambda$. Moreover, beside the meridional peaks at $\zeta = 1/c$ and $3/c$ already reported in the literature [259], corresponding to a chain periodicity $c = 10.3$ Å, a strong meridional peak at $\zeta = 5/c$ was revealed [260] (Fig. 31B).

In the analysis of [260], it was argued that the diffuse equatorial diffraction peak, centered at $\xi = 0.25$ Å^{-1} (Fig. 31A), gives information about the mean distance between chain axes, preferentially oriented parallel to the drawing direction, like in a nematic liquid crystal; the absence of diffraction peaks off the meridian was ascribed to a substantial absence of correlation in the position and orientation of the chains, besides their (imperfect) parallelism, whereas the substantial absence of layer lines (off the equator) was interpreted as due to the presence of conformational disorder.

The chain periodicity of PET in the mesomorphic form, $c = 10.3$ Å, is not far from the length of a monomeric unit in the highest extension. According to the results of conformational energy and geometrical analysis of [260], the conformational disorder compatible with the high extension of PET chains in the mesomorphic form would consist in random sequences of monomeric units in different minimum energy conformations, of the kind shown in Fig. 30B and C. Such low minimum energy conformers are characterized by all backbone dihedral angles in the *trans* state, except for CO – O – C – C bonds (Fig. 30B,C), which could assume in successive monomeric units along the chain, values close to + 80 and – 80° in addition to values close to 180°. A total of nine different minimum energy conformations were identified, for a single conformational repeating unit, which can randomly succeed each other, giving rise to a conformationally disordered chain straight and elongated. In the resulting conformationally disordered chains, the co-planarity of terephthalic groups along the chain, present in the normal triclinic form [258] (Fig. 30A), is lost in the long range (see Fig. 30B,C), while the average distance along c of the center of mass of consecutive monomeric units is close to the experimental value of 10.3 Å of the chain periodicity of PET in the mesomorphic form. The chains, although extended, present some waviness, which may account for the broadness of the equatorial halo, indicating the absence of defined "in plane" lattice periodicities of PET in the mesomorphic form.

Successively, this result was confirmed using Monte Carlo simulation techniques on model oligomers of PET in a confined geometry [261]. In these simulations, a model chain was placed inside a cylinder with hard walls (simulated by a continuous distribution of carbon atoms) in order to mimic the chain environment in the mesomorphic state. These calculations indicated that highly extended chain conformations of PET with transverse radius < 6 Å are compatible with narrow deviations from the *trans* state for all dihedral angles, with the exception of the backbone dihedral angles

CO – O – C – C, showing a flat distribution with broad maxima centered at 180, + 90 and – 90°.

Fourier transform calculations of isolated chain models have indicated that off the equator, the experimental diffraction pattern of the mesomorphic form of PET can be accounted for by extended chains obtained by random sequences of monomeric units in the above discussed minimum energy conformations [260]. This indicates a substantial absence of rotational (around the chain axes) and translational (parallel to the chain axes) order between adjacent parallel chains. The best agreement with experimental data was obtained for model chains characterized by a not co-planar arrangement of terephthalic residues along the chain, as a consequence of the presence of conformational disorder [260]. A typical result obtained by the average of the diffraction intensity calculated over at least 10 different conformationally disordered chain models, is shown in Fig. 32. Off the equator, the calculated patterns are in a good qualitative agreement with the experimental pattern of Fig. 31. Near the equator, the comparison between calculated and experimental diffraction intensity is less significant, because the interference between adjacent chains was completely neglected in the calculations of [260].

In conclusion, the mesomorphic form of PET corresponds to a solid mesophase of class iii), characterized by parallel arrangement of chain axes and long-range positional order of a structural feature only in one dimension, i.e. along the chain axis and absence of any long-range order in the lateral packing of the chains. Even though the chains in the metastable me-

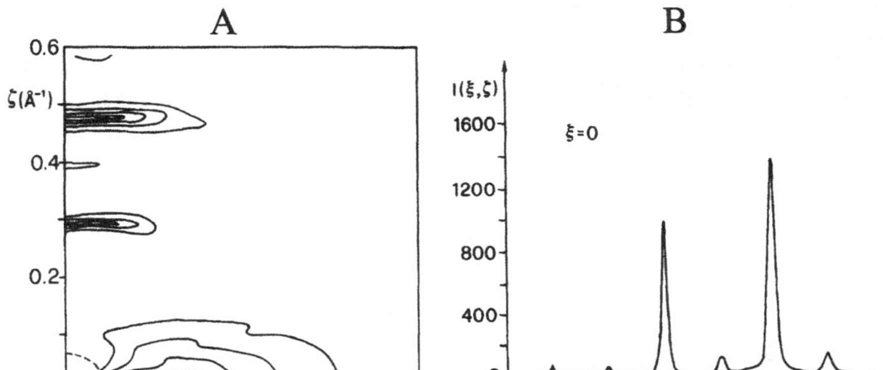

Fig. 32 (**A**) Calculated X-ray diffraction intensity $I(\xi,\zeta)$ for an irregular chain of PET of the kind shown in Fig. 30B and C as a function of the reciprocal cylindrical coordinates ξ and ζ, and (**B**) corresponding meridional profile. In (A), the contour lines correspond to values of $\log[I(\xi,\zeta)]$ equal to 3, 2.75, 2.5 and 2.25. (Reprinted with permission from [260]. Copyright 1992 by the American Chemical Society)

somorphic form of PET are in disordered conformations, monomeric units in different conformations follow each other along c at an average distance equal to 10.3 Å. Conformational disorder accounts for the absence of any diffraction along layer lines off the equator and off the meridian, and give a quite good agreement with the experimental data, as far as the distribution of X-ray diffraction intensity along the meridian. The packing of the parallel chains is completely disordered (mesophase of class iii)); in fact a lateral order in the relative position of chain axes, in the relative orientation of the chains around the chain axes and in the relative height of chains along the chain axes is maintained only in very short range, as indicated by the diffuse nature of the equatorial peaks in the experimental X-ray fiber diffraction pattern. The position of this equatorial peak, around $\xi \approx 0.25$ Å^{-1}, gives information concerning the average distance between structural motifs belonging to close neighboring chains, running in a nearly parallel relative arrangement.

6.2 Isotactic Polypropylene

Isotactic polypropylene (iPP), when quenched from the melt in cold water, gives rise to a metastable, solid mesophase, which transforms into the stable α form by annealing at elevated temperatures [25–27]. The X-ray powder diffraction pattern of this disordered form of iPP presents only broad halos at $2\theta = 14.8$ and $21°$ (Fig. 33A). Oriented fibers of iPP in this mesomorphic form can be obtained by stretching the corresponding unoriented films [25]. The X-ray fiber diffraction patterns present broad peaks on well-defined layer lines as shown in Fig. 33B; the chain periodicity corresponds to 6.5 Å, as in the crystalline α form.

On the basis of X-ray diffraction patterns and infrared spectra it was recognized that the partially ordered phase of iPP is made up of bundles of

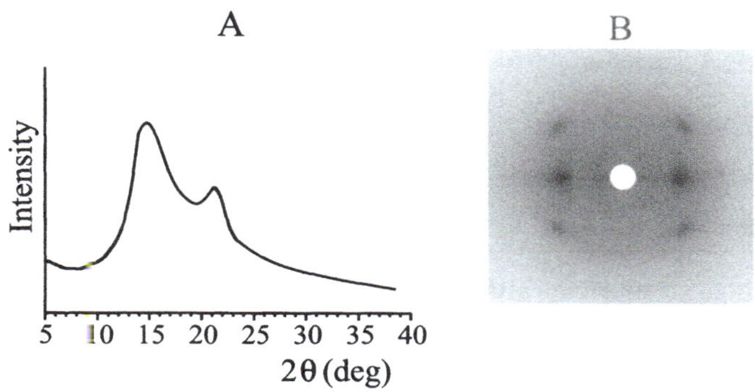

Fig. 33 X-ray diffraction patterns of iPP samples in the mesomorphic form: (**A**) powder pattern, (**B**) fiber pattern

parallel chains in the threefold helical conformation $(TG)_3$ as in the stable α form [25, 26]. It was also recognized that in the small mesomorphic aggregates long-range order is maintained only along the chain axes, whereas large amounts of disorder in the lateral packing of the helical stretches would be present [25, 26]. Owing to these structural features, Natta, Peraldo and Corradini [25] gave the name "smectic" to this mesophase, for which the X-ray diffraction data indicate short-range correlation in height among neighboring chains. This correlation corresponds to a degree of order higher than that of an ideal nematic liquid crystal, in which the only degree of order is the average parallelism of the rod-like molecules.

A comparison between calculated Fourier transform of various disordered models and experimental X-ray diffraction pattern have been reported in [262] and [263]. Structural models characterized by small bundles of parallel chains in 3/1 helical conformation packed with lateral disorder and keeping short-range correlations between neighboring chains similar to those of α form and the hexagonal β form of iPP, have been analyzed [262, 263]. Since the crystal structure of the β form had not been solved at that time, the hexagonal models built up for Fourier transform calculations in [262] and [263] were different from the now accepted structure of β form [264–266] even though the considered models had hexagonal correlations between chains.

The results of these analyses have indicated that the mesomorphic form of iPP is characterized by small bundles of parallel chains in ordered 3/1 helical conformation with disorder in the lateral packing [262]. The relative heights of neighboring chains within each bundle are mainly correlated. The local correlations are near to those which characterize the crystal structure of monoclinic α form [262, 263]. Any correlation about the relative position of atoms is lost at distances higher then 30–40 Å [262]. Since the lateral order is maintained at very short range, it was proposed that the term "paracrystalline" could be used for description of this form, if the paracrystalline distortion is intended to mean that a correlation exists within small nuclei of chains and fades away at longer interchain distances [262]

In conclusion, the metastable mesomorphic form of iPP, called at one time "smectic" [25], is a solid mesophase of class iii) characterized by long-range periodicity in one direction only, i.e. along the chain axes, due to the presence of chains in a 3/1 helical conformation.

6.3 Other Mesophases Characterized by Conformationally Ordered Polymer Chains and No Order in the Lateral Packing

Mesomorphic forms of the kind iii), characterized by conformationally ordered polymer chains and no order in the lateral packing, have been described for various isotactic and syndiotactic polymers. For instance, syndiotactic polypropylene (sPP) [28, 29, 267], syndiotactic polystyrene (sPS) [21, 22, 268, 269], syndiotactic poly(*p*-methylstyrene) (sPPMS) [270] and syndiotactic poly(*m*-methylstyrene) [271], mesomorphic forms have been described. In all of these cases the X-ray fiber diffraction patterns show diffraction confined in well-defined layer lines, indicating order in the conformation of the chains, but diffuse reflections on the equator and on the other layer lines, indicating the presence of disorder in the arrangement of the chain axes, as well as the absence of long-range lateral correlations between the chains.

The mesomorphic forms of sPP [29, 267], sPS [268, 269] and sPPMS [270] are characterized by chains in ordered *trans*-planar conformations packed in small disordered aggregates, which maintain the structural features of one of the polymorphic forms of the polymers, only up to very short distances. In the case of sPS and sPPMS, the small aggregates of chains are characterized by packing modes very close to those of the α form of sPS [268, 269] and form III of sPPMS [270].

7 Conclusions

In this paper, the structure of solid mesophases of polymers was discussed using the concept introduced by [13] that a solid mesophase may be intended as a state of the matter falling in between amorphous and the "ideal" crystalline states, which can be treated in terms of limit-disordered models implying maintenance of long-range positional order, at least along one dimension, of structural motifs which are not necessarily point centered. We have shown that the "not-point-centered" structural motifs may, for instance, consist of chain axes or a group of atoms along the chain.

Two classes of "solid mesophases" have been identified, class ii) and class iii). Generally, solid mesophases presenting conformationally disordered chains, but long-range order in the position of chain axes (condis crystals) belong to class ii), whereas solid mesophases presenting conformationally ordered chains, parallel among themselves, with disorder in the lateral packing, belong to class iii).

We have shown that this classification is sufficiently simple and general to include most important polymer "solid mesophases" identified so far. We are aware, however, that our classification of mesomorphic polymers, as well as alternative ones based on the different kinds of disorder which may occur in crystalline polymers, may not be exhaustive, since in many cases different kinds of disorder are interrelated and simultaneously present. For instance, in the case of polyacrylonitrile, the irregular configuration of chains determines conformational disorder in the crystalline state [97]. In the case of ethylene/propylene copolymers constitutional disorder implies also conformational disorder which, in turn, may induce rotational and translational disorder of the chains [195], because portions of chains close to the constitutional defect assume different orientation around the chain axes and different translation along the chain axes, in order to relieve local lattice strains. In the case of high-temperature forms of poly(tetrafluoroethylene), the presence of helix reversals necessarily implies that the helical portions of opposite chirality connected by the defect have a different orientation, thus generating rotational disorder [245]; at elevated temperatures, the lattice interactions become weaker, the mobility of the helix reversal defects increases, and the conformationally disordered chains appear poorly correlated as far as their relative heights and rotations around the z axis [245].

In the present article, the structure of only some of the many solid mesophases identified so far in the literature has been discussed in detail. Some other examples of solid mesophases are the high-temperature form of poly(diethyl siloxane) [272], alkoxy and aryloxy polyphosphazenes [273], and the normal crystal form of poly(methylene 1,3-cyclopentylene) [274, 275].

In particular, polymeric systems derived from the chemical modification of poly(tetrafluoroethylene) through the introduction of small concentrations of constitutional defects as comonomeric units have interesting structural and physical properties. For instance, random copolymers of tetrafluoroethylene with hexafluoropropylene (FEP), or with perfluoromethylvinylether (MFA), and perfluoropropylvinylether (PFA) are interesting because they present melt viscosities lower than that of poly(tetrafluoroethylene) and can be more easily processed as thermoplastic materials. All the random copolymers FEP, MFA and PFA crystallize at room temperature in the mesomorphic form I of PTFE (see Sect. 5.7), even at low concentrations of comonomeric units [276, 277]. The presence in the copolymers FEP, MFA and PFA of small amounts of constitutional defects $-CF_3$, $-OCF_3$ and $-OC_3F_7$, respectively, produces a high degree of disorder in the crystals and induces the formation of the mesomorphic form I of PTFE, already at room temperature [276, 277]. The intermediate form IV of PTFE is, indeed, not observed in these copolymers.

Let us finally remark the important role that solid mesophases play in manufacturing processes of polymeric materials. Polymers, in mesomorphic form, in fact, show enhanced mechanical properties and may be more easily processed, even at room temperature, than the stable crystalline forms. This allows obtaining manufactured objects at a low cost with high performances, which is not easy to obtain by processing the material directly in the stable crystalline state.

Acknowledgements Financial support from the "Ministero dell'Istruzione, dell'Università e della Ricerca" (PRIN 2002 and Cluster C26 projects) is gratefully acknowledged.

References

1. Friedel MG (1922) Ann Phys 18:273
2. Wunderlich B, Gerbowicz J (1984) Adv Polym Sci 60/61:1
3. Gray GW (1962) Molecular structure and the properties of liquid crystals. Academic, New York
4. Ciferri A, Krigbaum W, Meyer RB (1983) (eds) Polymer liquid crystals. Academic, New York
5. Gray GW, Winsor PA (1974) (eds) Liquid crystals and plastic crystals. Wiley, Chichester
6. Lehmann O (1907) Physik Z 8:42
7. Sherwood N (1979) The plastically crystalline state (Orientationally disordered crystals). Wiley, Chichester
8. Corradini P (1969) J Polym Sci Polym Lett 7:211
9. Suehiro K, Takayanagi MJ (1970) J Macromol Sci Phys B4:39
10. Bunn CW, Howells ER (1954) Nature 174:549
11. Clark ES, Muus LT (1962) Z Kristallogr 117:119
12. Corradini P, Guerra G (1977) Macromolecules 10:1410
13. Corradini P, Guerra G (1992) Adv Polym Sci 100:183
14. De Rosa C (2003) In: Green MM, Nolte RJM, Meijer EW (eds) Materials chirality: topics in stereochemistry. Wiley, Hoboken, NJ 24:71
15. Bassett DC, Khalifa BA, Turner B (1972) Nature 239:106
16. Bassett DC, Khalifa BA, Turner B (1972) Nature 240:146
17. Bassett DC, Block S, Piermarini GJ (1974) J Appl Phys 45:4146
18. Bonart R (1966) Kolloid Z 213:1
19. Bonart R (1966) Kolloid Z 210:16
20. Bonart R (1968) Kolloid Z 231:438
21. De Candia F, Filho AR, Vittoria V (1991) Macromol Chem Rapid Commun 12:295
22. Chatani Y, Shimane Y, Inoue Y, Inagaki T, Ishiota T, Ijitsu T, Yukinari T (1992) Polymer 33:488
23. Holmes DR, Bunn CW, Smith DJ (1955) J Polym Sci 17:159
24. Ziabicki A (1959) Kolloid Z 167:132
25. Natta G, Peraldo M, Corradini P (1959) Rend Fis Acc Lincei 26:14
26. Natta G, Corradini P (1960) Nuovo Cimento Suppl 15:40
27. Wyckoff HQ (1962) J Polym Sci 62:83
28. Nakaoki T, Ohira Y, Hayashi H, Horii F (1998) Macromolecules 31:2705
29. Vittoria V, Guadagno L, Comotti A, Simonutti R, Auriemma F, De Rosa C (2000) Macromolecules 33:6200
30. Bassi IW, Corradini P, Fagherazzi G, Valvassori A (1970) Eur Polym J 6:709
31. Baldwin FP, Ver Strate G (1972) Rubber Chem Technol 45:709
32. Walter ER, Reding EP (1956) J Polym Sci 21:561
33. Cole EA, Holmes DR (1956) J Polym Sci 46:245
34. Swann PR (1962) J Polym Sci 56:409
35. Wunderlich B, Poland D (1963) J Polym Sci 1:357
36. Baker CH, Mandelkern L (1965) Polymer 7:71
37. Crespi G, Valvassori A, Zamboni V, Flisi V (1973) Chim Ind (Milan) 55:130
38. Preedy J (1973) Polym J 5:13
39. Houtz RC (1950) Text Res 20:786
40. Bohn CR, Schaefgen JR, Statton WO (1961) J Polym Sci 55:531
41. Stefani R, Chevreton M, Garnier M, Eyraud C (1960) Hebd Seances Acad Sci

251:2174
42. Holland VF, Mitchell SB, Hunter WL, Lindenmeyer PH (1962) J Polym Sci 62:145
43. Klement JJ, Geil PH (1968) J Polym Sci Polym Phys Ed 6:138
44. Hinrichsen G, Orth H (1971) J Polym Sci Polym Lett Ed 9:529
45. Colvin BG, Storr P (1974) Eur Polym J 10:337
46. Gupta AK, Chand N (1979) Eur Polym J 15:899
47. Kumamaru F, Kajiyama T, Takayanagi M (1980) J Cryst Growth 48:202
48. Lindenmeyer PH, Hosemann R (1963) J Appl Phys 34:42
49. Pierce RHH Jr, Clark ES, Withney JF, Bryan WMD (1954) Proceedings of the 130th Meeting of the American Chemical Society, Atlantic City, NJ, 1954, p 9
50. Sperati CA, Starkweather HW Jr (1961) Fortschr Hochpolym Forsch 2:465
51. Kilian HG (1962) Kolloid Z Z Polym 185:13
52. Weeks JJ, Clark ES, Eby RK (1981) Polymer 22:1480
53. Farmer BL, Eby RK (1985) Polymer 26:1944
54. Yamamoto T, Hara T (1982) Polymer 23:521
55. Natta G, Corradini P, Porri L (1956) Rend Fis Acc Lincei 20:728
56. Natta G, Corradini P (1959) J Polym Sci 39:29
57. Moraglio G, Poliziotti G, Danusso F (1965) Eur Polym J 1:183
58. Corradini P (1975) J Polym Sci Polym Symp 51:1
59. Hosemann R (1963) J Appl Phys 34:25
60. Hosemann R, Bagchi SN (1962) (eds) Direct analysis of diffraction by matter. North Holland, Amsterdam
61. Welberry TR, Butler BD (1995) Chem Rev 95:2369
62. Giacovazzo C (2002) (ed) Fundamentals of crystallography, 2nd edn. IUCr series. Oxford University Press, Oxford
63. Guinier A (1994) X-ray diffraction in crystals, imperfect crystals and amorphous bodies. Dover, New York
64. James RW (1962) The optical principles of the diffraction of X-rays. Ox Pow, Woodbridge
65. Corradini P (1968) In: Ketley AD (ed) The stereochemistry of macromolecules. Marcel Dekker, New York 3:1
66. Natta G, Corradini P (1960) Nuovo Cimento 15:9
67. IUPAC Commission on Macromolecular Nomenclature (1979) Pure Appl Chem 51:1101
68. IUPAC Commission on Macromolecular Nomenclature (1981) Pure Appl Chem 53:733
69. Corradini P (1960) Rend Fis Acc Lincei 20:1
70. Huggins ML (1945) J Chem Phys 13:37
71. Bunn CW (1942) Proc R Soc Lond A180:67
72. Pauling L, Corey RB, Branson HR (1951) Proc Natl Acad Sci US 37:205
73. Corradini P (1969) Trans NY Acad Sci 31:215
74. IUPAC Commission on Macromolecular Nomenclature (1989) Pure Appl Chem 61:243
75. Hatada K, Kitayama T, Ute K, Nishiura T (2004) J Polym Sci Part A: Polym Chem 42:416
76. Turner Jones A (1965) Polymer 6:249
77. Natta G, Corradini P, Sianesi D, Moreno D (1961) J Polym Sci 51:527
78. Kakugo M (1995) Macromol Symp 89:545
79. De Rosa C, Talarico G, Caporaso L, Auriemma F, Galimberti M, Fusco O (1998) Macromolecules 31:9109

80. Tanaka A, Hozumi Y, Hatada K, Endo S, Fujishige R (1964) J Polym Sci B 2:181
81. Natta G, Porri L, Carbonaro A, Lugli G (1962) Makromol Chem 53:52
82. Guerra G, Di Dino G, Centore R, Petraccone V, Obrzut J, Karasz FE, MacKnight WJ (1989) Makromol Chem 190:2203
83. Auriemma F, De Rosa C, Corradini P (1993) Macromolecoles 26:5719
84. Auriemma F, De Rosa C, Boscato T, Corradini P (2001) Macromolecules 34:4815
85. VanderHart DL, Alamo RG, Nyden MR, Kim M-H, Madelkern L (2000) Macromolecules 33:6078
86. De Rosa C, Auriemma F, Perretta C (2004) Macromolecules 37:6843
87. Mooney BCL (1941) J Am Chem Soc 68:2828
88. Bunn CW (1948) Nature 161:929
89. Natta G, Corradini P (1956) J Polym Sci 20:251
90. Natta G, Bassi IW, Corradini P (1961) Rend Fis Acc Lincei 31:17
91. Smith RW, Wilkes CE (1967) J Polym Sci 5:433
92. Wilkes CE, Folt VL, Krimm S (1973) Macromolecules 6:235
93. Hobson RJ, Windle AH (1993) Macromol Chem Theory Simul 2:257
94. Flores A, Windle AH (1996)Comp Theor Polym Sci 6:67
95. Hinrichsen G, Orth H (1971) Kolloid-Z Z Polym 247:844
96. Liu XD, Ruland W (1993) Macromolecules 26:3030
97. Rizzo P, Auriemma F, Guerra G, Petraccone V, Corradini P (1996) Macromolecules 29:8852
98. Corradini P (1973) Ricerca Sci 84:3
99. Corradini P, Petraccone V (1972) (eds) Polymerization in biological systems. Ciba Foundation Symposium 7 (new series).
100. Natta G, Corradini P (1956) Angew Chem 68:615
101. Natta G, Corradini P (1960) Nuovo Cimento Suppl 15:111
102. Benedetti E, Corradini P, Pedone C (1975) Eur Polym J 11:585
103. Petraccone V, Ganis P, Corradini P, Montagnoli G (1972) Eur Polym J 8:99
104. McMahon PE, Mc Cullough RL, Schlegel AA (1967) J Appl Phys 38:4123
105. Petraccone V, Allegra G, Corradini P (1972) J Polym Sci Part C 38:419
106. Benedetti E, Ciajolo MR, Corradini P (1973) Eur Polym J 9:101
107. Predecki P, Statton WO (1966) J Appl Phys 37:4053
108. Reneker DH, Mazur J (1988) Polymer 29:3
109. Liang GL, Noid DW, Sumpter BG, Wunderlich B (1994) J Phys Chem 98:11739
110. Keith HD, Passaglia E (1968) J Res Natl Bur Stand 68A:513
111. Dreyfus P, Keller A (1970) Polym Lett 8:253
112. Warren BE (1990) X-ray diffraction. Dover, New York
113. Hikosaka M, Seto T (1973) Polym J 5:111
114. Natta G, Corradini P (1960) Nuovo Cimento Suppl. 15:40
115. Mencik Z (1972) J Macromol Sci Phys 6:101
116. Auriemma F, Ruiz de Ballesteros O, De Rosa C, Corradini P (2000) Macromolecules 33:8764
117. Guerra G, Petraccone V, Corradini P, De Rosa C, Napolitano R, Pirozzi B, Giunchi G (1984) J Polym Sci Polym Phys Ed 22:1029
118. De Rosa C, Guerra G, Napolitano R, Petraccone V, Pirozzi B (1984) Eur Polym J 20:937
119. De Rosa C, Guerra G, Napolitano R, Pirozzi B (1985) J Thermal Anal 30:1331
120. Corradini P, Giunchi G, Petraccone V, Pirozzi B, Vidal HM (1980) Gazz Chim Ital 110:443
121. De Rosa C, Corradini P, Buono A, Auriemma F, Grassi A, Altamura P (2003) Macro-

molecules 36:3789
122. Wunderlich B (1976) Macromolecular physics, vol 2. Academic, New York
123. Bassett DC (1976) Polymer 17:146
124. Keller A, Hikosaka M, Rastogi S, Toda A, Barham PJ, Goldbeck-Wood G (1994) J Mater Sci 29:2579
125. Keller A, Hikosaka M, Rastogi S, Toda A, Barham PJ, Goeldbeck-Wood G (1994) Philos Trans R Soc Lond A348:3
126. Rastogi S, Hikosaka M, Kawabata H, Keller A (1991) Macromolecules 24:6384
127. Lauritzen JI, Hoffman JD (1961) J Res Natl Bur Stand A64:73
128. Lauritzen JI, Hoffman JD (1961) J Res Natl Bur Stand A65:297
129. Hoffman JD, Davis GT, Lauritzen JI (1976) In: Hannay NB (ed) Treatise on solid-state chemistry, vol 3, chap 7. Plenum, New York
130. Sadler D (1983) Polymer 24:1401
131. Armitstead K, Goldbeck-Wood G (1992) Adv Polym Sci 100:221
132. Strobl G (2000) Eur Phys J E3:165
133. Lotz B (2000) Eur Phys J E3:185
134. Cheng SZD, Li CY, Zhu L (2000) Eur Phys J E3:195
135. Mathukumar M (2000) Eur Phys J E3:199
136. Warren BE (1941) Phys Rev 9:639
137. Hendicks S, Teller E (1942) J Chem Phys 10:147
138. Wilson AJC (1949) Acta Cryst 2:245
139. Wilson AJC (1952) Proc R Soc A180:277
140. Amoros JL, Amoros M (1968) Molecular crystals; their transforms and diffuse scattering. Wiley, New York
141. Hukins DWL (1981) X-ray diffraction of disordered and ordered systems. Pergamon, Oxford
142. Jagodzinski H (1987) Prog Cryst Growth Charact 14:47
143. Drits VA, Tchoubar C (1990) X-ray diffraction of disordered and ordered systems. Pergamon, Oxford
144. Mering J (1949) Acta Cryst 2:371
145. Brindley GW, Mering J (1951) Acta Cryst 4:441
146. Besson G, Tchoubar C, Mering J (1974) J Appl Cryst 7:345
147. Jagodzinski H (1949) Acta Cryst 2:201
148. Jagodzinski H (1954) Acta Cryst 7:17
149. Allegra G (1961) Nuovo Cimento 21:786; 22:661; (1961) Acta Cryst 14:535; (1964) Acta Cryst 17:579
150. Allegra G, Bassi IW (1980) Gazz Chim Ital 110:437
151. Giunchi G, Allegra G (1984) J Appl Cryst 17:172
152. Tadokoro H (1979) Structure of crystalline polymers. Wiley, New York
153. Kakudo M, Kasai N (1972) X-ray diffraction by polymers. Kodansha-Elsevier, Tokyo
154. Alexander LE (1979) X-ray diffraction in polymer science. Krieger, New York
155. Klug A, Franklin RE (1958) Discuss Farad Soc 25:104
156. Clark ES, Muus LT (1962) Z Kristallogr 117:108
157. Deas HD (1952) Acta Cryst 5:542
158. Ruland W, Tompa H (1968) Acta Cryst A24:93
159. Ruland W (1977) Koll-Z Z Polym 255:833
160. Lovell R, Windle AH (1976) Polymer 17:488
161. Cochran W, Crick FHC, Vand V (1952) Acta Cryst 5:581
162. Auriemma F, Petraccone V, Dal Poggetto F, De Rosa C, Guerra G, Manfredi C, Corradini P (1993) Macromolecules 26:3772

163. Auriemma F, Petraccone V, Parravicini L, Corradini P (1997) Macromolecules 30:7554
164. Iwayanagi S, Sakurai I, Sakurai T, Seto T (1968) J Macromol Sci Phys B2:163
165. Suehiro K, Takayanagi M (1970) J Macromol Sci Phys B4:39
166. De Rosa C, Napolitano R, Pirozzi B (1985) Polymer 26:2039
167. Schilling FC, Angeles Gómez M, Tonelli AE, Bovey FA, Woodward AE (1987) Macromolecules 20:2954
168. Boller A, Jin Y, Cheng J, Wunderlich B (1994) Thermochimica Acta 234:95
169. Iwayanagi S, Miura I (1965) Rept Progr Polym Phys Jpn 8:303
170. Ziabicki, A (1957) Collect Czech Chem Commun 22:64
171. Gianchandani J, Spruiell JE, Clark ES (1982) J Appl Polym Sci 27:3527
172. Arimoto H, Ishibashi M, Hirai M, Chatani Y (1964) J Polym Sci A 2:2283
173. Kinoshita Y (1959) Macromol Chem 33:1
174. Baldrian J, Lednicky F (1991) In: Puffr R, Kubanek V (eds) Lactam-based polyamides. CRC Press, Boston
175. Chakrabarti P, Dunitz JD (1982) Helv Chim Acta 65:1555
176. Li Y, Goddard III WA (2002) Macromolecules 35:8440
177. Natta G, Mazzanti G, Corradini P (1958) Rend Fis Acc Lincei 25:3
178. Henrici-Olivè G, Olivè S (1979) Adv Polym Sci 32:125
179. Hobson RJ, Windle AH (1993) Polymer 34:3582
180. Hobson RJ, Windle AH (1993) Macromolecules 26:6903
181. Stefani R, Chevreton M, Terrier J, Eyraud C (1959) Compt Rend 248:2006
182. Chiang J (1963) J Polym Sci A 1:2765
183. Chiang J (1965) J Polym Sci A 3:2109
184. Schaefer J (1975) Macromolecules 4:105
185. Kamide K, Yamazaki H, Okajiama K, Hikichi K (1985) Polym J 17:1291
186. Kamide K, Yamazaki H, Okajiama K, Hikichi K (1985) Polym J 17:1233
187. Kaji H, Schmidt-Rohr K (2000) Macromolecules 33:5169
188. Kaji H, Schmidt-Rohr K (2001) Macromolecules 34:7368
189. Kaji H, Schmidt-Rohr K (2001) Macromolecules 34:7382
190. Ver Strate G, Wilchinsky ZW (1972) J Polym Sci Part A2 9:127
191. Scholtens BJR, Riande E, Mark JE (1984) J Polym Sci Polym Phys Edn 22:1223
192. Mandelkern L (1964) Crystallization of polymers. McGraw-Hill, New York, pp 166–214
193. Eichorn RM (1958) J Polym Sci 31:197
194. Bunn CW (1939) Trans Faraday Soc 35:482
195. Ruiz de Ballesteros O, Auriemma F, Guerra G, Corradini P (1996) Macromolecules 29:7141
196. Seto T, Hara T, Tanaka K (1968) Jpn J Appl Phys 37:31
197. Muller A, Lonsdale K (1968) Acta Crystallogr 1:129
198. Treloar LRG (1975) The physics of rubber elasticity. Clarendon, Oxford
199. Kaminsky W, Bark A, Arndt M (1991) Macromol Chem Macromol Symp 47:83
200. Lang HT, Osan F, Wehrmeister T (1997) Polym Mater Sci Eng 76:22
201. Kaminsky W, Arndt-Rosenau M (2000) In: Scheirs J, Kaminsky W (eds) Matallocene-based polyolefins, vol. 2. Wiley, Chichester, p 91
202. Arndt M, Beulich I (1998) Macromol Chem Phys 199:1221
203. Harringhton BA, Crowher DJ (1998) J Mol Catal A: Chem 128:79
204. Cherdron H, Brekner M-J, Osan F (1994) Angew Makromol Chem 223:121
205. Grassi A, Maffei G, Milione S, Jordan RF (2001) Macromol Chem Phys 202:1239
206. Altamura P, Grassi A (2001) Macromolecules 34:9197

207. De Rosa C, Buono A, Auriemma F, Grassi A (2004) Macromolecules 37:9489
208. Jackson RL, Strage JH (1972) Acta Crystallogr B 28:1645
209. Nordman CE, Schmitkos DL (1965) Acta Crysallogr 18:765
210. Arndt M, Engehausen R, Kaminsky W, Zoumis K (1995) J Mol Catal A: Chem 101:171
211. Wunderlich B, Arakawa T (1964) J Polym Sci Part A 2:3697
212. Geil PH, Anderson FR, Wunderlich B, Arakawa T (1964) J Polym Sci Part A 2:3703
213. Arakawa T, Wunderlich B (1967) J Polym Sci Part C 16:653
214. Wunderlich B, Melillo L (1968) Macromol Chem 118:250
215. Keller A (1957) Philos Mag 2:1171
216. Wunderlich B (1976) Macromolecular physics, vol. 3. Academic, New York
217. Bassett DC, Khalifa BA (1973) Polymer 14:390
218. Yasuniwa M, Nakafuku C, Takemura T (1973) Polym J 4:526
219. Pechold W, Liska E, Grossman HP, Hägele PC (1976) Pure Appl Chem 96:127
220. Yamamoto T, Miyaji H, Asai K (1977) Japan J Appl Phys 16:1891
221. Yamamoto T (1979) J Macromol Sci Phys B16:487
222. Tanaka H, Takemura T (1980) Polymer J 12:355
223. Bassett DC, Turner B (1974) Phil Mag 29:225
224. Leute U, Dollhopf W (1980) Colloid Polym Sci 258:353
225. Bassett DC (1982) In: Bassett DC (ed) Development in crystalline polymers, vol 1, chap 2. Applied Science, London
226. Yasuniwa M, Enoshita R, Takemura T (1976) Jpn J Appl Phys 15:1421
227. Yasuniwa M, Takemura T (1974) Polymer 15:661
228. Ungar G (1986) Macromolecules 19:1317
229. Tashiro K, Sasaki S, Kobayashi M (1996) Macromolecules 29:7460
230. Ungar G, Keller A (1980) Polymer 21:1273
231. Clark ES (1973) Bull Am Phys Soc 18:317
232. (a) Clark ES (1999) Polymer 40:4659; (b) O' Leary K, Geil PH (1967) J Appl Phys 38:4169
233. Boerio FJ, Koenig JL (1971) J Chem Phys 54:3667
234. Johnson KW, Rabolt JF (1973) J Chem Phys 58:4536
235. Chantry GW, Fleming JW, Nicol EA, Willis HA, Cudby MEA, Boerio FJ (1974) Polymer 15:69
236. Yamamoto T, Hara T (1986) Polymer 27:986
237. Matsushige K, Enoshita R, Ide T, Yamauch N, Take S, Takemura T (1977) Jpn J Appl Phys 16:681
238. Hyndman D, Origlio GF (1960) J Appl Phys 31:1849
239. De Santis P, Giglio E, Liquori AM, Ripamonti A (1963) J Polym Sci Part B 1:1383
240. Brown RG (1964) J Chem Phys 40:2900
241. Clark ES (1967) J Macromol Sci Phys 1:795
242. Bates TW, Stockmayer WH (1968) Macromolecules 1:12
243. Bates TW, Stockmayer WH (1968) Macromolecules 1:17
244. D' Ilario L, Giglio E (1974) Acta Crystallogr Sct 30:372
245. Corradini P, De Rosa C, Guerra G, Petraccone V (1987) Macromolecules 20:3043
246. De Rosa C, Guerra G, Petraccone V, Centore R, Corradini P (1988) Macromolecules 21:1174
247. Kimming M, Strobl G, Stün B (1994) Macromolecules 27:2481
248. Weeks JJ, Sanchez IC, Eby RK, Poser CI (1980) Polymer 21:325
249. Wittmann JC, Smith P (1991) Nature 352:414
250. Wilson, FC, Starkweather HW Jr (1973) J Polym Sci Polym Phys Ed 11:919
251. Tanigami T, Yamaura T, Matsuzawa S, Ishikawa M, Mizoguchi K, Miyasaka K (1986)

Polymer 27:9[illegible]9; 27:1521
252. Iuliano M, De Rosa C, Guerra G, Petraccone V, Corradini P (1989) Makromol Chem 190:827
253. Petraccone V, De Rosa C, Guerra G, Iuliano M, Corradini P (1992) Polymer 33:22
254. Guerra G, De Rosa C, Iuliano M, Petraccone V, Corradini P, Ajroldi G (1993) Makromol Chem 194:389
255. D'Aniello C, De Rosa C, Guerra G, Petraccone V, Corradini P, Ajroldi G (1994) Polymer 36:967
256. De Rosa C, Guerra G, D'Aniello C, Petraccone V, Corradini P, Ajroldi G (1995) J Appl Polym Sci 56:271
257. Guerra G, De Rosa C, Iuliano M, Petraccone V, Corradini P, Pucciariello R, Villani V, Ajroldi G (1992) Makromol Chem 193:549
258. Daubeny RP, Bunn CW, Brown CJ (1954) Proc R Soc Lond 226:531
259. Asano T, Seto T (1973) Polymer J 5:72
260. Auriemma F, Corradini P, De Rosa C, Guerra G, Petraccone V, Bianchi R, Di Dino G (1992) Macromolecules 25:2490
261. Auriemma F, Corradini P, Guerra G, Vacatello M (1995) Mocromol Theo Simul 4:165
262. Corradini P, Petraccone V, De Rosa C, Guerra G (1986) Macromolecules 19:2699
263. Corradini P, De Rosa C, Guerra G, Petraccone V (1989) Polymer Commun 30:281
264. Meille SV, Ferro DR, Brückner S, Lovinger AJ, Padden FJ (1994) Macromolecules 27:2615
265. Lotz B, Kopp S, Dorset D (1994) C R Acad Sci Paris 319(II):187
266. Dorset DL, Mc Court MP, Kopp S, Schumacher M, Okihara T, Lotz B (1998) Polymer 39:6331
267. De Rosa C, Auriemma F, Ruiz de Ballesteros O, (2001) Polymer 42:9729; (2003) Polymer 44:6[illegible]79
268. Petraccone V, Auriemma F, Dal Poggetto F, De Rosa C, Guerra G, Corradini P (1993) Makromol Chem 194:1335
269. Auriemma F, Petraccone V, Dal Poggetto F, De Rosa C, Guerra G, Manfredi C, Corradini P (1993) Macromolecules 26:3772
270. Ruiz de Ballesteros O, Auriemma F, De Rosa C, Floridi G, Petraccone V (1998) Polymer 39:3523
271. De Rosa C, Buono A, Caporaso L, Petraccone V (2001) Macromolecules 34:7349
272. Beatty CL, Pochan JM, Froix MF, Hinman DD (1975) Macromolecules 8:547
273. Scneider NS, Desper CR, Beres JJ (1978) In: Blumstein (ed) Liquid crystalline order in polymers. Academic, New York
274. Ruiz de Ballesteros O, Venditto V, Auriemma F, Guerra G, Resconi L, Waymouth RM, Mogstad A (1995) Macromolecules 28:2383
275. Ruiz de Ballesteros O, Cavallo L, Auriemma F, Guerra G (1995) Macromolecules 28:7355
276. Centore R, De Rosa C, Guerra G, Petraccone V, Corradini P, Villani V (1988) Eur Polym J 24:445
277. Guerra G, Venditto V, Natale C, Rizzo P, De Rosa C (1998) Polymer 39:3205

Adv Polym Sci (2005) 181: 75–120
DOI 10.1007/b107175

Published online: 30 June 2005

Flow-induced mesophases in crystallizable polymers

Liangbin Li[1] (✉) · Wim H. de Jeu[1,2]

[1]FOM, Institute for Atomic and Molecular Physics,
Kruislaan 407, 1098 Amsterdam, The Netherlands
liangbin.li@Unilever.com, dejeu@amolf.nl

[2]Dutch Polymer Institute,
Eindhoven University of Technology, Eindhoven, The Netherlands
dejeu@amolf.nl

Abstract In this chapter we review flow-induced mesophases in some crystalline polymers with rather different characteristics: polyethylene terephthalate, isotactic polypropylene and polydiethylsiloxane. Because these three polymers are representatives of flexible and semi-rigid-chain polymers, the occurrence of a mesophase suggests a similar behaviour for many other crystalline polymers. The question may be not so much whether crystalline polymers show mesophase behaviour, but rather under which conditions it can be obtained. A flow field turns out to be an effective tool to unravel structural information, even if hidden in quiescent conditions. The emergence of a mesophase affects strongly any subsequent crystallization. As a result of templating and nucleating effects, a mesophase may accelerate the crystallization rate, lead to different crystal modifications, change the morphology and guide the orientation of the crystals. On the basis of mesophase formation new insights into flow-induced polymer crystallization emerge, even though the nature of the mesophase ordering is still not fully understood. A re-evaluation of models for the polymer melt (random coil and folded-chain fringed-micellar grains) is proposed to understand the combined experimental observations on mesophase ordering and crystallization.

Keywords · Mesophase · Crystallization · Random coil · Folded-chain fringed-micellar grains · Flow

1 Introduction

Flow-induced phase transitions are fundamental properties of non-equilibrium systems. As moreover flow fields are inevitably applied during polymer processing, the ordering processes under such conditions have attracted increasing interest [1]. Knowledge of flow-induced crystallization in the polymer melt is essential to control the final properties of the products from industrial processing methods such as extrusion, injection, and blow moulding. Nevertheless, no satisfactory molecular theory exists that can describe polymer crystallization under flow conditions [2, 3]. During the past half a century, a great deal of phenomenological studies has been dedicated to this subject; see for a recent review [4]. On this basis several aspects of flow-induced crystallization of polymers are well established [4, 5]. (i) Flow accelerates the crystallization kinetics and – under severe conditions – changes the semi-crystalline morphology from spherulites to crystallites oriented in the flow direction. (ii) The enhancement of the crystallization kinetics is attributed to an increase in the nucleation rate caused by distortions of the polymer chains in the melt. (iii) The molecular weight and the molecular weight distribution both have a pronounced influence. This phenomenological description of flow-induced crystallization leaves an essential question still open: How does flow promote nucleation in polymer crystallization?

Even at quiescent conditions the fundamental mechanisms of polymer crystallization, especially at an early stage, are still poorly understood [6–16]. For many years, nucleation and growth as a stepwise process has dominated the discussion [11, 12]. In contrast to this view Strobl [13] proposed a multi-stage process to explain polymer crystallization, while others concluded on the basis of X-ray scattering data to a spinodal-assisted process [17–29]. Common to both views is that the crystallization is preceded by an ordered precursor (so-called pre-ordering). Clear structural information about such possible precursors – necessary to verify these hypotheses – is still scarce. As a result, during recent years an important and still open debate has been going on about polymer crystallization.

Interestingly, pre-ordering was already implied in some rather early studies of polymer crystallization, but did not get the present type of attention. As early as 1967 Katayama et al. [30] observed a small-angle X-ray scattering (SAXS) peak significantly earlier than the corresponding crystalline Bragg peaks in wide-angle X-ray scattering (WAXS). They proposed that density fluctuations occurred before the formation of any crystals. The idea of a multi-stage process dates back to 1967 by Yeh and Geil [31, 32], while

Schultz introduced in 1981 a spinodal approach promoting orientation in polymer systems [33]. However, the essential question about the nature of pre-ordering still remained open. Moreover, the description 'pre-ordering before crystallization' is often not used in a precise way. A polymer system cannot develop such ordering because it wants to crystallize. The large enthalpy associated with crystallization/melting prevents pre-transitional fluctuations as found for second-order or weakly first-order phase transitions. Hence any pre-ordering before crystallization in polymers is necessarily an independent effect that enables subsequent crystallization. Evidently, the precursor should possess some ordering intermediate between the liquid and the crystal phase.

Recently, several flow-induced mesophases have been reported, some of which were not observed at quiescent conditions. Because clear structural information is obtained, on this basis a better understanding and possibly answers to the above questions can be extracted. This is the main motivation to for this review. In this section we shall first give an introduction to polymer mesophases and their possible influence on crystallization. In Sect. 2 flow-induced mesophases are reviewed for three representative polymers: polyethylene terephthalate (PET), isotatic polypropylene (iPP) and polydiethylsiloxane (PDES). The molecular structure of these polymers is depicted in Scheme 1. Because of the semi-rigid or flexible nature of the chains, they are in principles not expected to show liquid crystallinity. Hence induced rigidity is an essential element for the formation of a mesophase [34–36]. The influence of the mesophase on subsequent crystallization is discussed in Sect. 3. The analysis indicates that the answers to fundamental questions regarding polymer ordering may be found in the disordered (molten) state. This leads us in Sect. 4 to an evaluation of the limitations of the random coil model for the polymer melt and amorphous state to capture local ordering. The review ends with some concluding remarks. Note that we use the term 'flow field' in a generalized way, which includes stretching, drawing and shearing. We restrict ourselves to mesophase ordering in bulk homopolymers, block copolymers are not included.

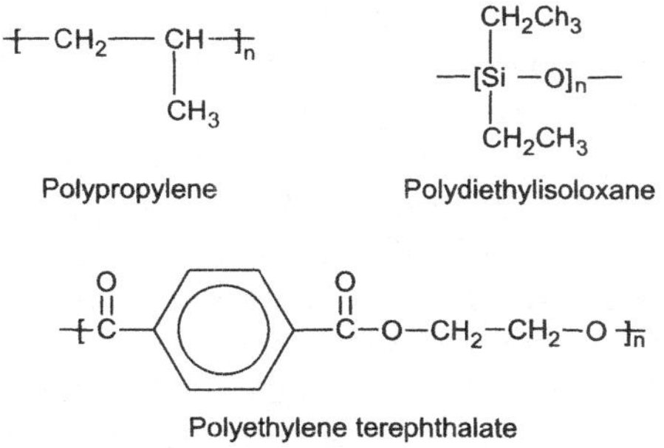

Scheme 1

1.1 Polymer mesophases

The different states of condensed matter can be characterized by the various types of molecular organisation: positional order and orientational order. The latter aspect is directly related to the molecular mobility. Depending on these criteria, liquid phases, mesophases, crystalline phases, amorphous glasses and mesoglasses can be distinguished [37–41]. In the field of low-molecular-weight liquid crystals the terms 'mesophase' and 'liquid crystal (LC) phase' are identical [37, 38]. The crucial notion common to all LC phases is long-range orientational order of the preferred axis of the molecules. Restricting us for simplicity to uniaxially symmetric molecules, the preferred axis can be the long axis (rod-shaped molecules) or the short axis (disc-shaped molecules). For nematic phases this is all; there is no further positional order (anisotropic liquid). Smectic and columnar LC phases possess in addition reduced positional order, forming either a layered structure (for rod-like molecules) or a columnar structure (consisting of disc-shaped molecules). These mesophases can be formed as a function of temperature (thermotropic liquid crystals) or in solution (lyotropic mesophases from surfactant-like molecules). Note that the alternative possibility of full positional order and no orientational order (plastic crystals of solid-rotator phase) does not fall in the class of LC phases, as the molecules form in spite of their rotational freedom a full three-dimensional lattice. The possible extension to glassy states from either of these phases with reduced mobility is obvious.

In low-molecular-mass systems the rod- or disk-like molecules (mesogens) often contain aromatic rings (giving a certain stiffness) connected via more flexible groups and usually also with flexible end groups [37, 38]. Evidently these elements can also be present in polymers. If this is explicitly the case and various nematic, smectic and columnar phases are observed as a function of temperature, we speak of main-chain LC polymers [42, 43]. Alternatively mesogens can be attached via flexible spacers to a polymer backbone: side-chain LC polymers that will not be of our concern. The macromolecular equivalent of lyotropic (surfactant) LC phases is provided by block copolymers [44]. In macromolecular science the concept of a mesophase is often used in a rather general sense, in agreement with the original meaning of 'intermediate phase'. Apart from LC phases also conformationally disordered or condis crystals are incorporated in this category, even though they largely maintain positional and orientational order [39, 40]. The situation is further complicated because in the field of polymers the term mesophase does not necessarily refer to a homogeneous thermodynamically stable phase, similarly as for crystallized polymer structures.

The concept of a condis crystal was suggested in 1975 [45] and their existence was documented by many examples in 1984 [39]. Condis crystals have in general an hexagonal packing and some authors have stressed the analogy with columnar liquid crystals [46]. They show positional and orientational order of the molecule as a whole, but are partially or fully conformationally disordered and mobile. Especially linear, flexible macromolecules can display a high chain mobility introducing large-amplitude conformational motions around the chain axis [40]. Nevertheless the position and orientation of the molecule is left unchanged. Typical examples are the high-temperature crystal phases of polyethylene (PE), polytetraflouoroethylene (PTFE), trans-1,4-polybutadiene (PB), and PDES. They resemble the low-temperature crystal phases of LC forming molecules with long flexible end groups [47].

Starting point to discuss mesophases in polymers is the concept of orientational ordering. Evidently this requires some form of stiffness and LC theories are often based upon rigid rods. One of the most successful ones is the Flory lattice model [48], which builds on the more general Onsager theory [49]. In this model the free energy of the LC system depends only on $\psi L/D$, the product of the volume fraction ψ and the aspect ratio L/D of the rods. Flory's lattice model predicts a critical value $\psi L/D = 6.4$ for an isotropic-nematic transition. For semi-rigid or wormlike chains, the length L of the rigid rod is replaced by the Kuhn length l_k. In the LC theory of Maier and Saupe [50] the rigidity of molecules is not explicitly stressed. Instead it is included in an orientation-dependent interaction that comprises exclude-volume effects and van der Waals attraction. In polymers rigidity can be explicitly built in via mesogenic units that can perform a condensation reaction. When rigidity is not directly present in the individual chains, it can be induced by self-assembly – as found in many biopolymers [51] – or by conformational changes like a coil-helix transition [34–36]. In addition to LC ordering, many flexible chains have the possibility to form condis crystals on the basis of rigidity-induced by conformational ordering [39, 40]. The original definition of a condis crystal emphasized disordered conformations, using the normal crystal as reference state. On the other hand, compared to a disordered liquid the conformations of chains in condis crystals are partly ordered as otherwise no ordering could exist at all.

In polymer systems with (semi-)flexible chains, mesophase ordering is based on rigidity induced by conformational ordering. Using phenomenological arguments, de Genes and Pincus showed that the coupling between a coil-helix transition and orientational ordering could induce a nematic phase [34, 35]. The conformational state of the macromolecule is determined by two parameters [52]: (1) An activity coefficient $s = \exp(-\Delta F_{hc}/k_B T)$ where ΔF_{hc} is the difference in free energy per monomer between a randomly coiled (disordered) and a rigid (ordered) configuration, $k_B T$ is the Boltzmann factor. (2) A co-operativity parameter $\sigma = \exp(-2\Delta F_g/k_B T)$ where ΔF_g is the free energy necessary to create a boundary between rigid and coiled segments.

The fraction f_h of units in the helicoidal state is given by

$$f_h = \frac{1}{2} + \frac{(s-1)}{2\sqrt{(s-1)^2 + 4\sigma s}} \,. \tag{1}$$

In this situation $s < 1$ means that the coil state is energetically favoured. De Gennes and Pincus [34] introduced a Maier–Saupe type interaction [50] between the rigid (helical) segments, which leads an additional force favouring helical domains. In a mean-field approximation the activity factor can be written as $\tilde{s} = hs$, in which h is a coupling constant originating from the mean field. In the case of excluded-volume effects [35] the mean-field treatment always gives $h > 1$. This means that the helical state can be energetically favourable even for $s < 1$. If the excluded-volume interactions are sufficiently strong, a phase transition can occur at which the coupling drives a first-order coil-helix transition while simultaneously long-range orientational correlations develop. Flory and Matheson reached similar conclusions based upon the lattice model [36].

1.2
Flow-induced mesophases and its influence on crystallization

The highly asymmetric nature of long-chain polymer molecules creates quite generally possibilities for the formation of mesophases with orientational order. However, for the overwhelming majority of flexible-chain polymers regions of spontaneously ordered segments are too small to manifest anisotropic properties. This situation changes dramatically upon application of a flow field that promotes alignment of the molecular chains. Though at quiescent conditions the overall shape of the molecular chains in the melt as well as in the amorphous state is spherically symmetric, the aspect ratio L/D can reach values like 10 000 in a completely extended state. This tremendous change of aspect ratio induced by an external flow field provides considerable opportunities to tailor the mesophase behaviour of polymers.

In LC systems with intrinsic rigid building blocks, flow suppresses fluctuations and shifts the phase transition to higher temperatures or lower concentrations [53–56]. With respect to flexible chains, present theoretical work on the formation of flow-induced LC structures does not include the possible effect of flow on conformational ordering. The analysis of experimental data under flow are usually based upon a generic polymer model of homogeneous chains stretched in their entropic regime, like freely jointed [57, 58] or wormlike [59, 60] chain models. A considerable amount of experimental and theoretical work has been published on coil-helix transitions induced by external fields in a single chain system [61–65]. However, to our best knowledge coupling between conformational and orientational ordering in multi-chain systems under flow has not been considered so far. Buhot and Halperin performed a mean-field treatment of the helix-coil transition induced by stretching a single chain [63] to which Tamashiro and Pincus added a correction [65]. In this approach, an elastic contribution is added to the Hamiltonian in order to incorporate the effect of the externally applied force f. The analysis shows a plateau in the force-extension profile (stress-strain curve), which indicates the coil-helix transition. In principle, the concept of induced rigidity [34–36] leading to mesophase ordering of flexible chains under flow, could be included in this approach. However, the actual situation is more complicated than just adding another effect. Moreover Buhot and Halperin's arguments are still based upon a thermodynamic equilibrium approach [63]. Flow-induced conformational ordering modifies the rigidity of the building blocks of the polymer chains and can transform a flexible chain into a rigid one. Consequently, apart from possibly shifting the isotropic-mesophase transition temperature as for rigid-chain polymers, flow can also lead to new phases. In analogy to a theoretical study of the coupling between adsorption and the helix-coil transition [66], a rich phase diagram is expected under flow conditions. The coupling between conformational and orientational ordering under non-thermodynamic (flow) condition remains an important theoretical challenge. It is crucial for understanding flow-induced mesophase ordering and crystallization.

Brochard and de Gennes [67] discussed theoretically a flow-induced isotropic-mesophase transition in a polydisperse polymer system occurring through spinodal decomposition. Following Maier–Saupe's [50] theory of the nematic phase, the orientation-dependent interaction energy was taken as

$$U_{ij} =- C(\alpha_{\parallel} - \alpha_{\perp})^2 S_i S_j \,, \tag{2}$$

where $\alpha_{\parallel}$ and $\alpha_{\perp}$ represent the polarizability of a molecule parallel and perpendicular to the preferred axis, $S_i = 1/2(3\cos^2\theta_i - 1)$ is the orientation parameter and C is a constant. Based on this orientation-dependent interaction energy, Brochard et al. [67] formulated a Flory-Huggins-type free energy F for a bimodal homopolymer system as

$$\frac{F}{k_B T} = \frac{\phi_L}{N_L} \ln\phi_L + \frac{\phi_S}{N_S} \ln\phi_S - \chi \sum_{ij} \phi_i \phi_j S_i S_j \,. \tag{3}$$

Here $\phi_{L,S}$ and $N_{L,S}$ represent the volume fraction and the monomer number of the long and short chains, respectively. The Flory–Huggins parameter χ is defined as $\chi = \frac{z}{2} C(\alpha_{\parallel} - \alpha_{\perp})^2$ in which z is the coordination number of the monomer. Upon imposing a flow field, S_i varies with chain length, which subsequently leads to an instability resulting in phase separation between oriented (long) and non-oriented (short) chains. The thermodynamic instability appears at a zero of the second-order derivative of the free energy $F''(\phi_L)$, which results in

$$\frac{1}{N_L\phi_L} + \frac{1}{N_S\phi_L} = \chi [S_L(t) - S_S(t)] \,. \tag{4}$$

Here $S_L(t)$ and $S_S(t)$ are expressed as $S_i(t) = S_o M_i(t)$, where $M_i(t)$ is the memory function describing the relaxation of molecular chains after having been subjected to a flow field:

$$M_i(t) = 0.8 \exp(-t/\tau_i), \quad (t > 0.1\tau_i)\,. \tag{5}$$

For $N_L > N_S > N_e$ (N_e is the critical entanglement molecular weight), τ_i is the reptation time, while for $N_S \approx N_e$ it represents the Rouse time [60, 69]. Because τ_i has a strong molecular weight dependence, long chains keep their orientation while short already relax to the isotropic state. For $N_L \gg N_S$ one finds [70]

$$S_o = 1/5 N_e^{-1}(\lambda^2 - 1/\lambda)\,, \tag{6}$$

where λ is the extension ratio. Inserting Eqs. 5 and 6 in Eq. 4, we obtained a critical extension ratio λ_c for spinodal decomposition:

$$\lambda_c = \frac{5N_e}{(2\phi_S N_S \chi)^{1/2}}, \quad N_L \gg N_S\,. \tag{7}$$

These results indicate that spinodal decomposition can be induced by flow even in a homopolymer system, which results in an oriented nematic phase with long chains and an isotropic liquid with short chains. Though the above argument is based on a bimodal system, the same principle has also been applied to the polydisperse case [67]. However, this approach still does not take flow-induced conformational ordering into account, which may couple to the anisotropic interactions.

In spite of the clear analogy between stress-induced mesophase formation and stress-induced crystallization under flow, there is a significant difference between the two processes. Thermodynamically, for a phase transition accompanied by a change in length ΔL, the relationship between the force f and the temperature T is given by [71, 72]

$$[\partial(f/T)/\partial(1/T)]_P = \Delta H/\Delta L\,, \tag{8}$$

where ΔH is the melting enthalpy of either the mesophase or the crystal. When $\Delta H > 0$, usually $\Delta L < 0$ and f/T will increase with T. This means that at constant pressure the phase transition temperature will increase for an increase in the applied tensile force. Equation 8 was originally intended for isotropization of a stretched mesophase network as well as for melting of a stretched crystalline network. However, the enthalpy of isotropization of a mesophase is at least one order of magnitude smaller than that of melting of crystallites [73]. Therefore, the force coefficient of the transition temperature for isotropization of a mesophase, defined as dT/df, should be at least one order of magnitude larger in comparison to the melting of a stretched crystalline phase. On this basis, we can expect a large number of order-disorder transitions in polymers, hidden below the crystallization temperature at quiescent conditions, to be unveiled by a flow field.

The central question in this review is how shear (or stretching or drawing) influences possible mesophase formation. The mesophase, in turn, can strongly influence the crystallization process. Obviously, crystals with intrinsically parallel chains cannot form spontaneously without some prior orientation of the chains, either locally or over larger distances. Hence any form of orientational order of the chains will have an enabling effect on the crystallization process. According to Ostwald's stage rule [74], a phase transformation from one stable state to another one proceeds via metastable states – whenever these exist – in stages of increasing stability. Such states are not only intermediate forms on the path to the final state but can also provide a key to understanding the crystallization behaviour [75–80]. Computer simulation shows that in a system with a metastable liquid-liquid phase coexistence line, critical density fluctuations enhance the nucleation rate dramatically around the critical temperature [81]. Olmsted et al. [24] presented a phase diagram of a homopolymer that possesses this feature. Comparing with the situation at either higher or lower temperatures, an increase of the nucleation rate by a factor of 10^{13} is estimated around the critical point, while the growth rate does not change appreciably [81]. However, in semi-rigid or flexible-chain polymers, crystallization generally occurs before reaching the phase separation or mesophase ordering temperature because of its metastability at quiescent condition. A flow field can switch the relative stability of mesophases and crystals due to an order of magnitude difference in sensitivity to the external stimuli, which ensures that polymer chains to reach conditions equivalent to those around a critical point.

In addition to providing nucleation sites, a mesophase can also serve as a template with an intermediate order during the transition from a disordered melt to a fully ordered crystal. Because the molecular chains adapt in the mesophase already a certain orientational, positional and conformational ordered state, kinetically it will be relatively easy to transform from a mesophase to a crystalline state.

2
Flow-induced mesophases

In this section, we review three representative cases of flow-induced mesophases. PET serves as an example of a semi-rigid polymer with intrinsic rigid building blocks. The shear-induced smectic ordering in the flexible-chain polymer iPP provides a special case of induced rigidity, which is still far from fully understood. PDES has been chosen as an example of a stretch-induced condis crystal. Various experimental techniques have been employed to characterize flow-induced mesophases. While scattering methods probe the ordering, a direct technique to measure the molecular mobility is nuclear magnetic resonance (NMR). Up to now, work combining NMR and a flow field is limited [82, 83], while X-ray scattering has been employed extensively in combination with shear, spinning and drawing. Hence most experimental results presented in this review are based on X-ray scattering. The reader is referred to the original publications and to a recent review [84] for experimental details.

2.1 Draw-induced mesophase in polyethylene terephthalate

The crystallization and mesophase behaviour of PET provides a good example for semi-rigid chain polymers including poly(ethylene nathphalate) (PEN) and poly(ether ketone ketone) (PEKK), etc. [85]. Upon crystallization, PET forms a triclinic unit cell with $a = 0.456$ nm, $b = 0.594$ nm, $c = 1.075$ nm, $\alpha = 98.5°$, $\beta = 118°$, and $\gamma = 112°$ [86]. The melting point T_{m} and the glass transition temperature T_{g} are about 256 °C and 80 °C, respectively. The crystallization properties of PET have been well investigated [87, 88], both at quiescent conditions and under external fields such as flow and high pressure. However, most of the experimental work under flow was restricted to drawing at temperatures around the glass transition.

Under quiescent conditions PET normally does not form a mesophase. Following the criteria for the formation of a LC phase of stiff polymers of Doi et al. [60, 89, 90], Imai et al. [20] estimated that neither in the glass nor in the melt a LC phase can appear in PET. Gonzalez et al. reported a persistence length $l_{\mathrm{p}} = 1.2$ nm for a PET molecule in the theta state [91]. From the value of the van der Waals width of a benzene ring, the diameter of a PET 'rod' is estimated to be $d \approx 0.66$ nm. Inserting these values in the equation

$$\nu^* = \frac{16}{\pi d l_{\mathrm{p}}^2}, \tag{9}$$

the critical concentration of rods for an isotropic-nematic transition is calculated to be $\nu^* = 5.4$ segments/nm^3. The value estimated in the glass state is 4.2 segments/nm^3, resulting in $\nu < \nu^*$ and no liquid crystallinity is expected. Nevertheless, upon annealing at temperatures close to T_{g}, density fluctuations were observed before the onset of crystallization. During the induction period a structure with a length scale of about 20 nm developed following spinodal decomposition kinetics. Yeh et al. observed such a process in real space [31, 32]. In this context it is interesting that transmission electron microscopy (TEM) revealed in amorphous PET ball-like structures with a diameter of 4.5–10 nm. During annealing around T_{g}, diffusion and aggregation of the 'balls' was observed leading subsequently to a lamellar structure. This process can be fitted into Strobl's multi-stage approach to crystallisation [13] and may also explain the spinodal kinetics measured by

SAXS [17–29]. Though the structures were suggested to be nematic, direct evidence is absent.

Bonart was the first to report that drawing induced mesophases in PET [92, 93]. During stretching of a totally amorphous PET sample he observed first the formation of a nematic phase and subsequently a smectic one. Asano et al. investigated the structural changes during annealing of cold-drawn amorphous PET films by X-ray diffraction and micro-indentation [94, 95]. Figure 1 shows two-dimensional (2D) X-ray scattering at angles between SAXS and WAXS of cold-drawn PET during annealing at different temperatures [95]. The drawing speed and the average draw ratio were 0.8 mm/min and 3.8, respectively. A weak but sharp reflection appeared on the meridian at temperatures from 60 to 80 °C, assigned to smectic layering with a spacing of 1.07 nm and denoted as (001′). Upon increasing the temperature to 70 °C, crystallization started slowly and the smectic phase disappeared at about 100 °C. The mechanical properties of the smectic mesophase were different from those of both the oriented amorphous and the crystalline phase. At 20 °C the density of the isotropic amorphous phase, the oriented glass phase, the smectic phase and the crystalline phase were reported to be 1.335, 1.37, 1.38 and 1.455 g/cm^3, respectively. Hence a sharp increase in density occurred only upon crystallization.

Figure 2a summarizes X-ray scattering results of oriented amorphous PET during annealing at different temperatures [95], which points to a structural development as pictured in Fig. 2b. The meridional reflection is first apparent at an annealing temperature of 50–60 °C (depending on annealing time). The spacing is almost equal to the monomer unit length. This suggests that below T_g neighbouring molecular segments tend to match each other laterally.

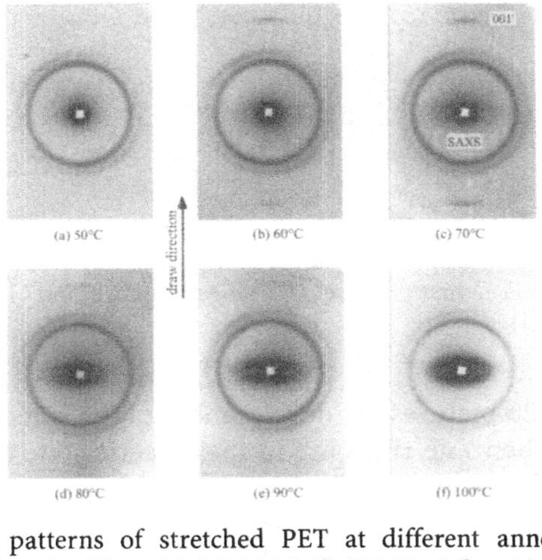

Fig. 1 In-situ x-ray patterns of stretched PET at different annealing temperatures; (**a**) 50 °C (**b**) 60 °C (**c**) 70 °C (**d**) 80 °C (**e**) 90 °C (**f**) 100 °C (after Asano et al. [95])

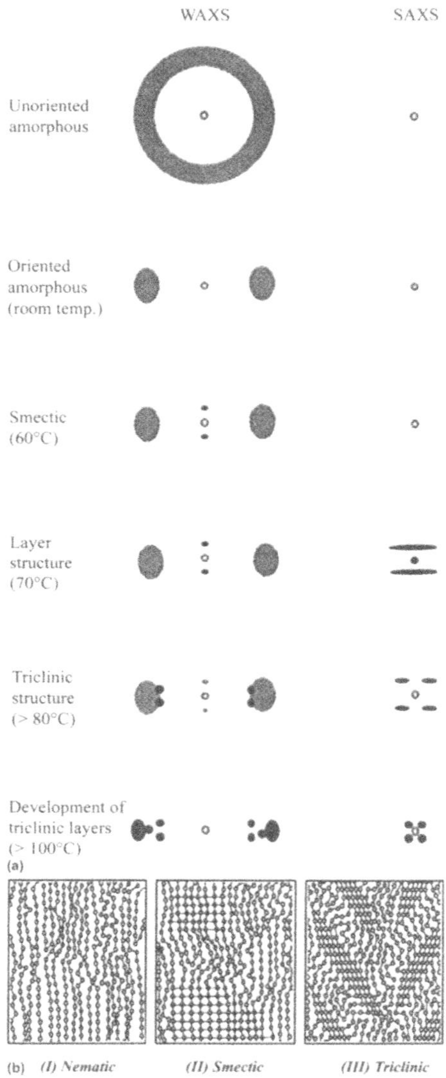

Fig. 2 Crystallization of stretched PET. (**a**) Summary of x-ray results (**b**) Cartoon picture of the morphologies. (I) nematic phase, (II) smectic phase and (III) triclinic structure (after Asano et al. [95])

Frame II from Fig. 2b pictures the development of the smectic phase from the nematic state. The benzene rings are in the smectic state arranged in planes perpendicular to the draw direction, whereas the lateral packing of neighbouring molecules shows no positional order. The smectic phase could be attained through an increase in the mobility of the molecular segments leading to a slightly more densely packed structure ($\varrho = 1.38\,\mathrm{g/cm^3}$) than in the nematic state.

At a constant temperature below T_g, a critical draw strain is required to induce the smectic phase [96, 97]. Figure 3 shows WAXS patterns on the meridian during cold draw at 50 °C for different draw strains. The smectic phase appears at a draw strain of about 58%. In addition to the smectic peak at $2\theta = 8°$ two additional peaks are observed at about 16° and 26° (X-ray wavelength of 0.154 nm). The peak at 16° is weak and not mentioned in the original report; it could be the second order of the (001′). The peak at 26° was assigned as (− 103). This reflection lies neither on the crystallographic c^*-axis nor on the axis normal to the (a^*b^*)-plane. For triclinic symmetry the latter does not coincide with the c^*-axis and corresponds to the c-axis in real space.

The appearance of the smectic phase is influenced by temperature as well as by draw rate [98–103]. At 90 °C, the smectic ordering can be induced by a draw rate of 13 $\mathrm{s^{-1}}$, and no meridional smectic diffraction can be resolved below 10 $\mathrm{s^{-1}}$. Blundell et al. observed that the smectic phase appears about 250 ms before the onset of crystallization (measured with a time resolution of 40 ms/frame). Kawakami et al. [104, 105] confirmed that at a temperature of 90 °C no smectic peak appears at low draw rates. However, combining X-ray scattering and the strain-stress relation (see Fig. 4), they concluded that a nematic or smectic phase is induced by draw in zone I. The draw-induced phase transition from the melt to the mesophase is first-order. PET with a low average molecular weight shows a similar phase behaviour but the transition zone I (mesophase) is narrower than for higher average molecular weight.

The groups of Blundell, Hsiao, Keum, and Fukao [106] reported a smectic spacing of about 1.03 nm, somewhat smaller than both the 1.07 nm originally given by Asano et al. [94] and the c-axis repeat of the triclinic crystal phase (1.075 nm) [86]. Two different proposals have been put forward for the chain packing in this case. Ran et al. attributed the short spacing to a smectic-C structure [96]. In that case the chain axes are tilted with respect to the layer

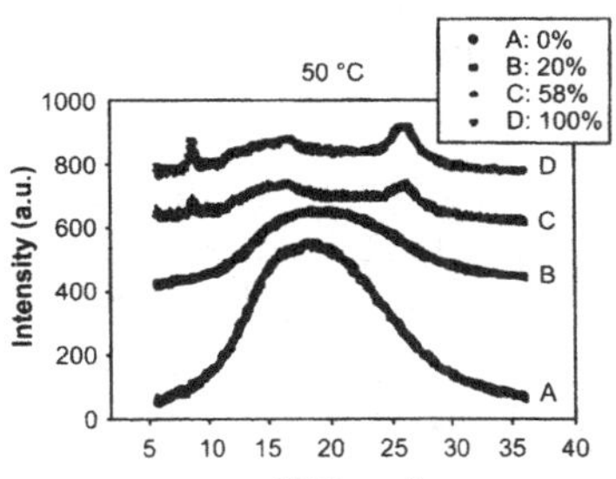

Fig. 3 One-dimensional x-ray intensity profiles of a PET film at different deformation stages: strain 0%, 20%, 58% and 100% (Calculated from WAXS on the meridian by azimuthal averaging from 85 °C to 95 °C)(after Ran et al. [96])

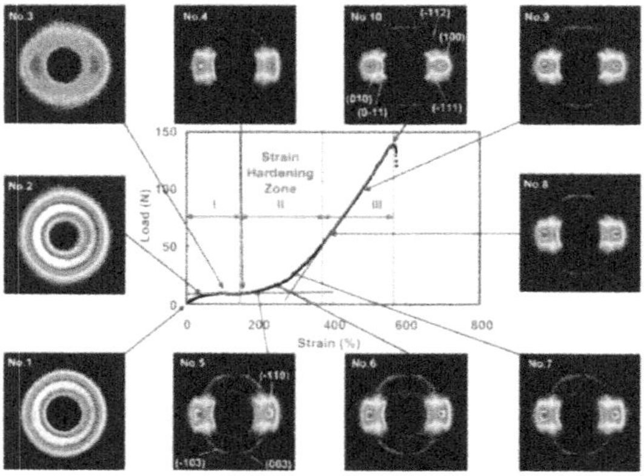

Fig. 4 Load-strain relation and selected WAXS patterns during stretching of high molecular weight PET (after Kawakami et al. [104])

normal, which is still parallel to the stretching direction (Fig. 5a). In such a model the layer peak should split into two reflections off-axis. This has not been observed, which is not fully conclusive as it could be due to overlap with the broad mesophase peak on the equator. We remark that this interpretation does not consider the possible occurrence of conformational disorder in the polymer chains. An alternative model assumes that the chains are parallel to the stretching direction and form perpendicular layers, but that the layer spacing is reduced by conformational changes (Fig. 5b) [107]. Analysis of the energy associated with various geometries suggests that in the smectic phase the PET chains should be nearly in all-trans conformations. However, the CO – O – CC dihedral angle could be close to 80°. This arrangement leads to a layer spacing shorter than the crystalline *c*-axis. The random sequences of monomeric units generate chains without any long-range correlation between the planes of the phenyl rings along the chains. In our opinion the second model is more plausible. Apart from lack of support from the X-ray scattering for the smectic-C model, a systematic tilt of the chains with respect to the draw direction seems unlikely.

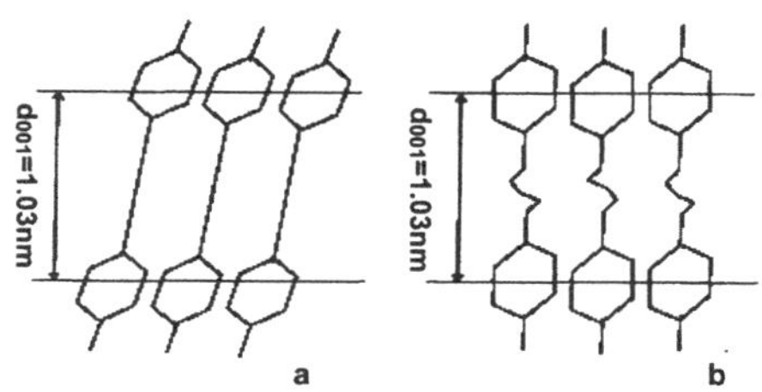

Fig. 5 Two possible packing models for the mesophase in PET (**a**) smectic-C model (**b**) conformational disorder model

As mentioned earlier, the persistence length of PET chains is slightly smaller than needed for an isotropic-mesophase transition. Upon stretching and annealing amorphous PET at temperatures close to T_g, conformational ordering is expected to promote rigidity. The combination of intrinsic rigidity and induced rigidity by conformational ordering [34–36] can easily meet the criterion for an isotropic-mesophase transition. In addition, a flow field enhances the spontaneous alignment of the molecular chains. Kinetically the resulting orientational ordering could even follow a spinodal model at a mean field situation [67].

Finally we mention that draw-induced mesophases have also been observed in some other semi-rigid chain polymers. Examples of such studies are cold drawing of glassy PEN [108, 109] and oriented crystallization of PET/PEN copolymers [110, 111]. In all cases a smectic mesophase has been reported. In the absence of crystallization it can attain a degree of stability that resists decay from chain relaxation. This supports the view that the mesophase represents a thermodynamic state [101]. We expect that similar mesophases based on semi-rigid monomers can be observed in other polymers with intrinsic rigid building blocks, such as other aromatic polyesters and polyamides.

2.2
Shear-induced smectic phase in isotactic polypropylene

Isotactic polypropylene is a typical semi-crystalline flexible-chain polymer. Three crystal modifications α, β and γ have been found, which all show a three-fold helical conformation of the chains [112]. A 'mesosphase' has been reported in rapidly quenched samples [113]. Though its nature used to be a matter of controversy, it is now generally accepted that it is metastable and readily transforms into a crystalline phase at temperatures above 80 °C [114–120, 122, ?–126]. Due to its representativity, stability and commercial importance, during the past half a century a great deal of work has been dedicated to flow-induced ordering of iPP. However, because in a flexible-chain polymer like iPP no liquid crystalline phase is expected at high temperatures, the focus has been on shear-induced crystallization rather than on low-dimensional ordering [127–150]. Recently, we reported that a smectic phase is induced in iPP by shearing; this phase is found at temperatures above as well as below the melting point [6, 151, 152]. In this respect iPP seems to be rather unique. A flow-induced smectic phase has been suggested in polydimethylsiloxane (PDMS), but direct evidence is still absent [83].

Figure 6a gives a typical SAXS picture of the shear-induced smectic peak at 200 °C; an accompanying smectic WAXS peak may be hidden in the amorphous halo of the corresponding pattern in Fig. 6b [152]. At this temperature no crystallization is expected at all, as confirmed by the absence of any crystalline reflections in both SAXS and WAXS. The smectic SAXS intensity is

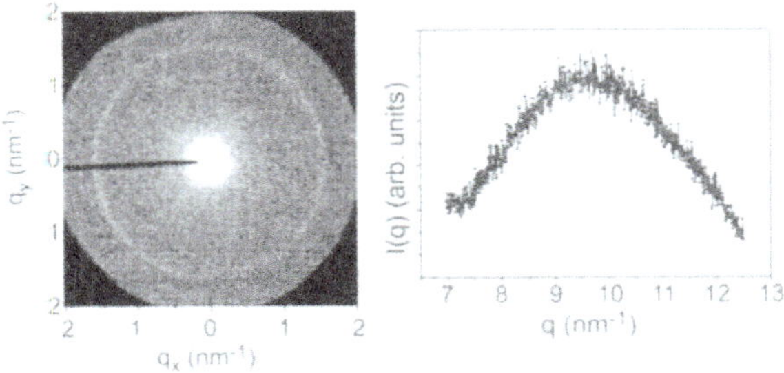

Fig. 6 (**a**) Two-dimensional SAXS pattern of iPP displaying smectic ordering at 200 °C (**b**) and corresponding one-dimensional WAXS pattern after a steady shear with shear rate 1 s^{-1} for 2 min (after Li and de Jeu [152])

weak, due to small volume fraction of the smectic domains and the limited density contrast of the order of 7%. The smectic peak is sharp, giving a correlation length of the order of several tens of nanometer. The smectic phase can be obtained at a wide temperature range up to 225 °C by applying the same shear field. It disappears at 225 °C after about 1 hour but can survive longer than overnight at 170 °C. At temperatures below 170 °C the shear-induced smectic ordering is followed by crystallization [150]. In this situation both conventional crystalline lamellae with a large periodicity (about 25 nm and crystallized smectic layers (with an approximately 10% increased periodicity of about 4.5 nm) are present. In the following we shall refer to the latter layering still as the smectic peak, even though the underlying structure may have changed with temperature and crystallization time.

The nature of the smectic layering was confirmed by a shear alignment [6]. Figure 7a shows the weak initial smectic peak in the SAXS pattern, slightly oriented, as obtained by a shear pulse at 180 °C. After imposing a steady shear with a shear rate of 0.1 s^{-1} for 10 min, the smectic peak disappears completely (Fig. 7b). Rotating the sample over 90° aligns the X-ray beam parallel to $\boldsymbol{v} \times \dot{\boldsymbol{v}}$

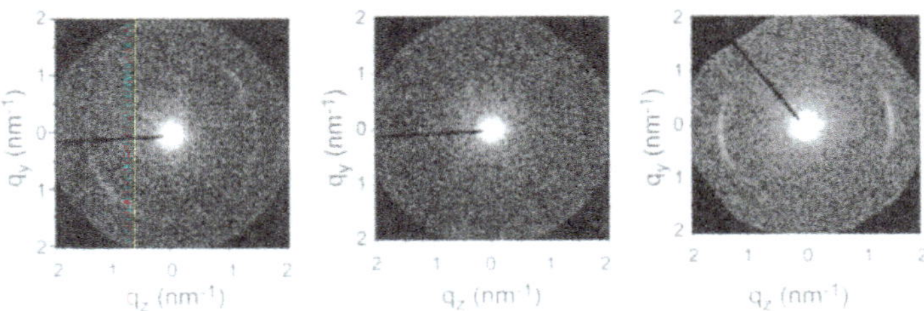

Fig. 7 (**a**) Two-dimensional SAXS pattern showing a weakly oriented smectic peak for an iPP sampla subjected to a shear pulse with a rate of 1 s^{-1} and a strain of 3000% at 180 °C (**b**) Disappearance of the smectic peak after a steady shear of 0.1 s^{-1} for 10 minutes (**c**) SAXS pattern with the incident x-ray beam parallel to $\boldsymbol{v} \times \dot{\boldsymbol{v}}$ showing highly oriented smectic layers (after Li and de Jeu [6])

and a highly oriented smectic peak is visible again (Fig. 7c). These alignment effects make it difficult to observe the smectic layering in a two-plate shear configuration, even though it can already be induced by pressing or any other small deformation during sample preparation.

The formation of the smectic phase in iPP is attributed to induced rigidity [34–36]. In an ordered state a molecular chain of iPP adapts in general a 3/1-helix conformation [112]. In the melt IR measurements indicate still short helical sequences up to about 5 monomers. However, upon shearing at 180 °C this number increases up to 14 (see Fig. 8). Potentially this can supply the stiffness for an isotropic to mesogenic transition. Upon increasing the temperature to 190 °C, IR bands corresponding to helical sequences with up to 12 monomers still survive. As X-ray scattering indicates that only the smectic ordering survives above 180 °C, these helical sequences with up to 12–14 monomers must be the building blocks of the smectic layers.

According to Doi et al. [29, 60, 89, 90, 153] the criterion for an isotropic to nematic transition is a critical persistence length

$$l_{\mathrm{p}} = \frac{4.19 M_0}{b l_0 \varrho N_{\mathrm{A}}} \,. \tag{10}$$

In this equation b represents the diameter of a polymer segment, ϱ the density, N_{A} Avogadro's number, and M_0 and l_0 the mass and the length of a monomer, respectively. For the iPP melt, $M_0 = 42\,\mathrm{g/mol}$, $b = 0.665\,\mathrm{nm}$, $l_0 = 0.217\,\mathrm{nm}$ and $\varrho = 0.85\,\mathrm{g/cm^3}$. Inserting these values into Eq. 8 we find $l_{\mathrm{p}} \approx 2.38\,\mathrm{nm}$, which is for a 3/1 helix equivalent to about 11 monomers [152]. The rod length of iPP helices as obtained from IR measurements meets this criterion (see Fig. 9). However, present theories predict a nematic phase rather than a smectic one [34–36, 89, 90]. A broad distribution of the helical sequence length should exist in the melt of flexible polymers in a dynamic equilibrium, which contributes a favourable entropy term. If the conformational ordering and orientational ordering occur in sequence, only a nematic

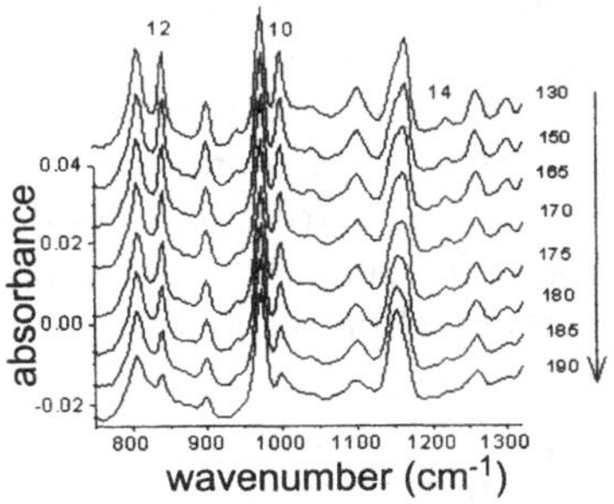

Fig. 8 IR spectra of a sheared iPP sample during a heating scan. The monomer number associated with the helical sequence of a specific band is indicated (after Li and de Jeu [6])

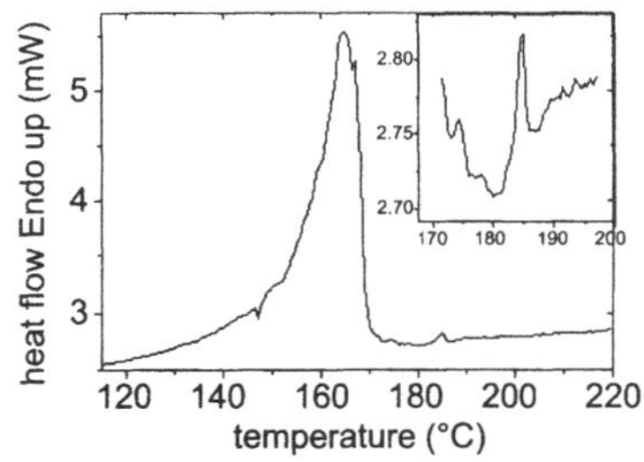

Fig. 9 DSC curves of a melt-sheared iPP sample during a heating scan with a rate of 10 °C/min (after Li and de Jeu [6])

phase would be expected. The selection of the sequence lengths happens during the formation of the smectic layers, thus the conformational ordering (or coil-helix transition) and the alignment of helices take place simultaneously. The coupling between conformational ordering and density determines an average period consisting of a rigid helical sequence of at least 11 monomers (about 2.6 nm) and a random-coil part of about 1.4 nm. As is well known from low-molecular weight smectic phases, in mixtures the smectic layers form a weight average of the long dimensions of the respective components [154, 155]. If we accept that the same applies to a polydisperse distribution of helical sequences and random-coil parts in iPP, the average helical/random-coil dimension comes in as a natural length scale determining the smectic periodicity.

Upon heating a crystallized sample the smectic regions show a higher melting temperature than their crystalline counterpart. Figure 9 displays a typical DSC curve of a sheared iPP1 sample during heating. In addition to the large melting peak of iPP crystals around 165 °C, several additional small melting peaks appear at higher temperatures. The insert in Fig. 9 shows a blow-up of a DSC trace from 170 to 195 °C. We attribute these high-temperature melting peaks to contributions from different smectic regions. Polarized optic microscopy (POM) observation show that the smectic layers assemble into fibrillar entities, which melt in the same temperature range as found for the melting of the smectic layers by X-ray scattering, DSC and FTIR.

The fibrillar entities have been modelled as anisotropic drops containing smectic domains with a size of tens of nanometers (see Fig. 10). After imposing a shear pulse, the smectic layer normal follows the direction of the long axis of the anisotropic drops (Fig. 10a). After shearing for a longer time, the smectic layers rotate and finally align in the direction of the velocity gradient (see Fig. 7). In this model, the smectic layers reorient within the fibrillar entities that keep their long axes in the original direction following the flow (Fig. 10b). The relative orientation of the smectic layers, the crystalline lamellae and the fibrillar entities, revealed by X-ray scattering in Fig. 11, lies at the origin of this model. The two-dimensional SAXS pattern of Fig. 11 is from

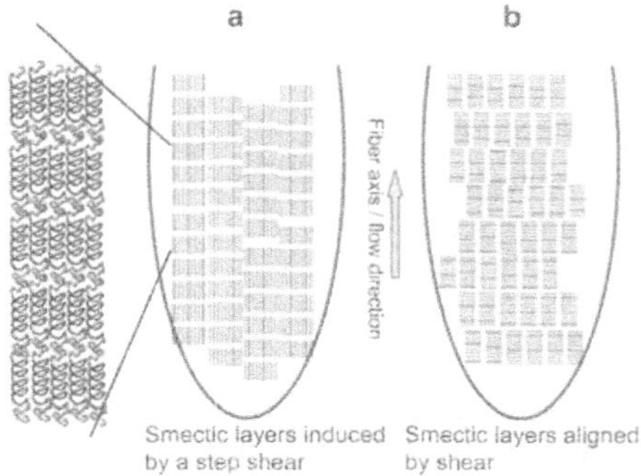

Fig. 10 Anisotropic drop model of smectic domains in iPP induced by shear (**a**) at initial shear alignment (**b**) after full shear alignment (after Li and de Jeu [6])

a crystallized iPP sample with initially pre-aligned smectic layers above the melting point. In addition to the scattering peaks from the oriented smectic layers and the crystalline lamella, two additional strikes are observed close to the beamstop. These are contributions from the anisotropic drops. The smectic layers are parallel to the long axis of the anisotropic drops and to the normal of the crystalline lamellae. The latter two follow the flow direction (see arrow in Fig. 11), as observed in all flow-induced crystallization experiments. The anisotropic drop model provides a conserved system and thus can explain the reversible changes of the smectic periodicity during heating and cooling [6].

The molecular weight and its distribution have a significant influence on the smectic ordering in iPP. Figure 12a gives one-dimensional SAXS curves at 190 °C of three iPP materials that differ in this respect. The smectic spacing from the different materials is plotted vs temperature in Fig. 12b. In all cases the smectic periodicity decreases with increasing temperature. Comparing results at the same temperatures, the smectic layers of iPP with a higher average molecular weight have a larger periodicity, while a larger molecular

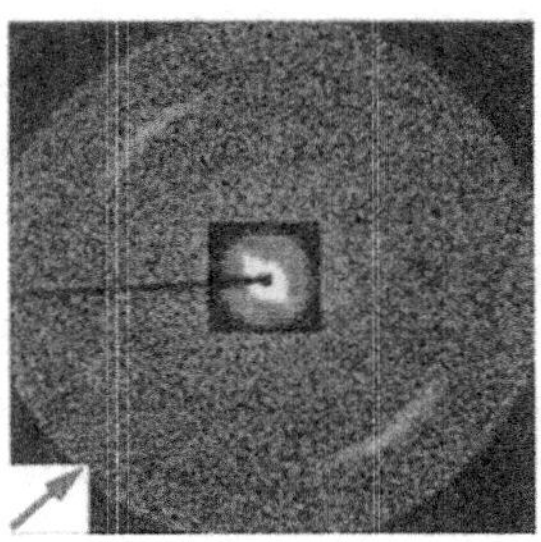

Fig. 11 Two-dimensional SAXS pattern showing at small angles the perpendicular direction of the smectic layers with respect to the bundle axis (after Li and de Jeu [6])

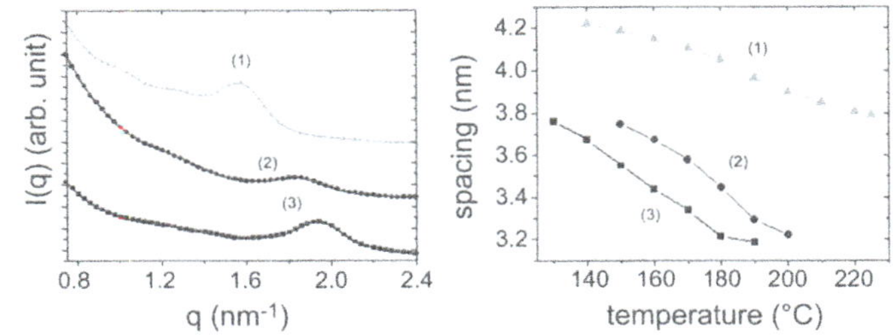

Fig. 12 (**a**) Typical one-dimensional SAXS pattern of three sheared iPP materials with different molecular characteristics at 190 °C (**b**) Smectic periodicity vs temperature For iPP 1, 2 and 3 the average molecular weights M_w are 720 k, 580 k and 370 k and their distributions M_w/M_n: 4.8, 3.6 and 3.7, respectively (after Li and de Jeu [6])

weight distribution gives a larger intensity of the smectic peak. Because the smectic layer spacing and the length of a polymeric chain are at completely different length scales, a direct connection is not expected. According to arguments of Brochard et al. [67] about flow-induced spinodal decomposition in polydisperse homopolymer systems, the molecular weight distribution (or the ratio between the length of long and short chains) is an important factor to affect the formation of any mesophase. A larger chain length difference leads to a higher concentration of smectic regions to which long chains are supposed to give the main contribution.

Let us next consider smectic layering in iPP due to fibre spinning which imposes an elongation flow [6, 156]. With the X-ray beam perpendicular to the fibre axis – the usual geometry for in-situ X-ray scattering on fibres – a crystalline peak rather than smectic layering is observed. The smectic peak appears only after positioning the X-ray beam parallel to the fibre axis. This indicates that the smectic layers are oriented parallel to the fibre axis. In principle this averaging of an already small peak gives a rather low scattering that can easily be missed.

The 'mesophase' observed in the past in fast-quenched iPP is probably a transient phase that occurs when cooling the high-temperature smectic phase to crystalline. In the melt or supercooled melt before crystallization, the molecular chains in the smectic layers are still in a liquid state. During the crystallization process of the smectic layers, two broad WAXS peaks are observed, similar as reported for quenched samples and used as evidence for the low-temperature 'mesophase' [113–120, 122, ?–126]. After some time these two peaks disappear. This strongly suggests that the structure observed in quenched iPP is a transient metastable state during the transition from the (high-temperature) smectic phase to crystalline. Through a fast quenching this transient phase can be easily frozen in. Somewhat different observations by various authors [113–120, 122, ?–126] can be attributed to differences in thermal history of their samples, which changes the state of the high-temperature smectic phase used as starting point for quenching.

2.3 Stretch-induced condis crystal in polydiethylsiloxane

Polydialkylsiloxanes belong to a group of non-polar, rather flexible macromolecules that are known to form mesophases in spite of the fact that they do not contain typically mesogenic units or amphiphilic groups [82, 157–176]. PDES is the first member of the polydialkylsiloxanes, which are capable of forming two mesophases denoted as μ_1 and μ_2 [172]. With respect to conformational disorder, these mesophases have been described as condis phases [39, 40], regarding 2D ordering they can be compared to columnar discotic phases [46]. Figure 13 presents a schematic picture of the molecular packing in these phases. Upon decreasing the temperature from the mesophases, PDES can crystallize into two crystalline phases, α (monoclinic) and β (tetragonal) [163, 164]. The thermodynamically more stable β_1 and β_2 forms develop preferably upon slow cooling from the mesophases μ_1 and μ_2, while a high fraction of α-PDES is obtained upon fast cooling. Still another crystal structure of PDES has been described and designated as a γ-modification [164]. It closely resembles the tetragonal β polymorphs, and is thermodynamically less stable than the α phase. Figure 14 shows a schematic phase diagram of PDES [172]. The transition temperature from mesophase to melt is strongly influenced by molecular weight, while mesophase-crystal and crystal-crystal transition temperatures are relatively weakly affected. For a molecular weight < 28 000, no mesophase is expected to form at all [163]. The strong molecular weight dependence of the mesophase-melt transition is still not fully understood.

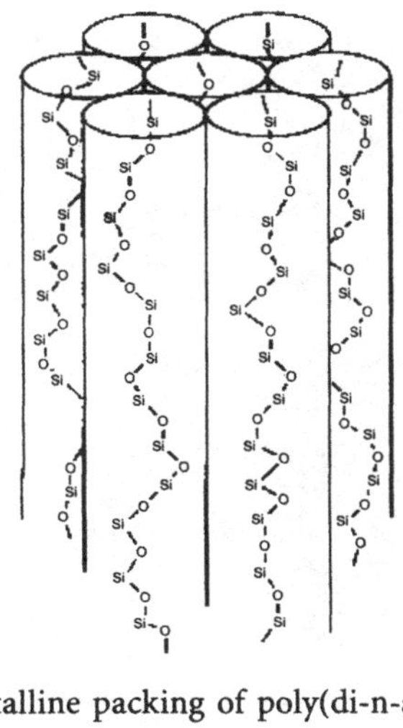

Fig. 13 Columnar liquid-crystalline packing of poly(di-n-alkylsiloxane) molecules (after Molenberg et al. [164])

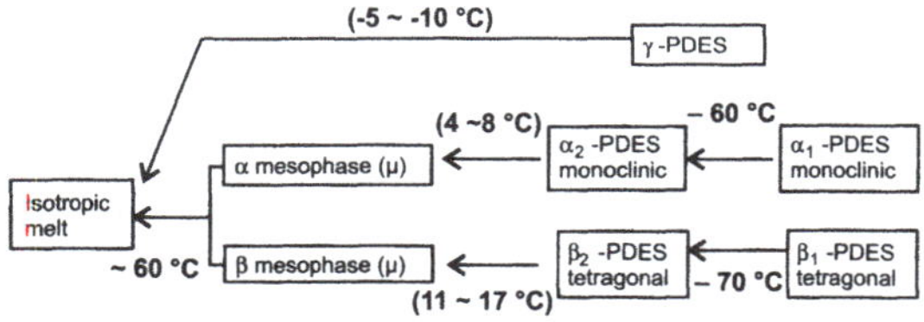

Fig. 14 Phase diagram of PDES (after Ganicz and Stanczyk [172])

Cross-linked networks of PDES have been investigated in some detail, and are also capable of forming $\alpha, \beta, \gamma, \mu$ and amorphous phases. In Godovski's pioneering work the main aim was to obtain well-oriented crosslinked samples by mechanical stretching in order to compare the phase behaviour of unoriented and oriented PDES [167]. A striking phenomenon exhibited by amorphous PDES networks is a stress-induced transition to the mesophase. This stretch-induced isotropic-mesophase transition is illustrated by microscopic images in Fig. 15 [82]. When amorphous PDES elastomers are stretched to a critical extension ratio λ (generally $\lambda > 2.0$), a neck region forms. After a few minutes the neck region stabilizes under the stress and assumes a well-defined shape with a distinct transition region (Fig. 15a). The neck is highly birefringent while at several places a faint crosshatched fibrous structure is visible. When the stress is released gradually, the neck is incorporated in the amorphous regions starting from the transition zone (Fig. 15b).

Fig. 16 gives a schematic diagram of the so-called engineering stress (force divided by the amorphous/isotropic cross section in the absence of any force) *vs* extension ratio for an initially amorphous PDES elastomer [82]. Upon

a

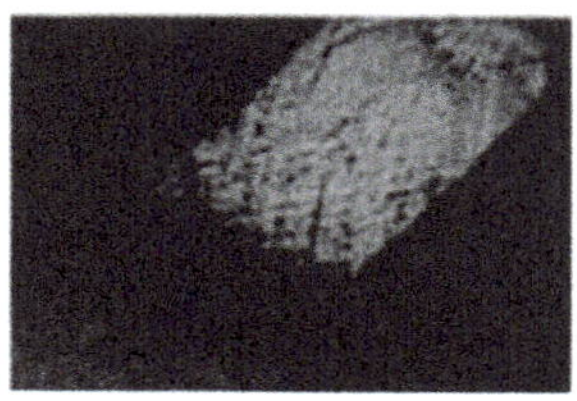

b

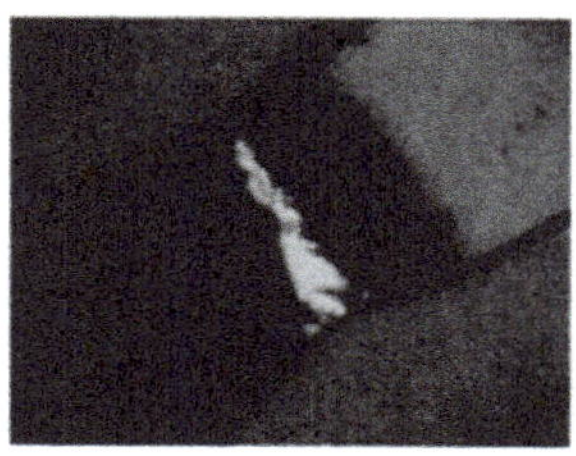

Fig. 15 POM pictures of the neck region of PDES (**a**) after stabilization (**b**) when the stress is reduced (after Hedden et al. [82])

stretching the network, it lengthens from point O to point A while remaining in the amorphous phase. Upon further stretching from point A to point B in the metastable amorphous region, no phase transition is apparent, partly because the mesophase formation is kinetically limited. If the sample is held long enough at B, a neck forms and the engineering stress decreases to point C, indicating mesophase formation. A differential increase in the engineering stress is required to stretch the sample from point C to point D. During this process, amorphous regions are converted into the neck until the entire sample is necked at point D. Further stretching toward the breaking point E requires a significant increase of the engineering stress because the network chains are already highly extended. If the stress is released slowly from a sample initially at point D, it follows the path D $\rightarrow$ C $\rightarrow$ A $\rightarrow$ O without entering into the metastable amorphous region. Within the neck region, a larger extension ratio corresponds to a higher concentration of the mesophase [82].

The amorphous-mesophase transition induced by stretch is a 2D growth process [174]. Figure 17 shows WAXS patterns at different times during the transition from amorphous to mesophase of a stretched cross-linked PDES network. Initially a broad peak due to scattering from the amorphous region dominates the pattern. At later times a sharp mesophase peak emerges at the expense of the broad amorphous one. The two peaks are close, and at intermediate times the amorphous peak appears as a broad shoulder at the small-angle side of the mesophase peak. The oriented amorphous region has a spacing of 0.81 nm while that of the mesophase is 0.79 nm. From the development of the scattered intensity of the mesophase, the kinetics of the amorphous-mesophase transition can be described by an Avrami model with an exponent $n = 1.98$. This suggests that the transition can be characterised as a 2D growth process.

Upon increasing the stress, the transition temperature of amorphous-mesophase increases (Fig. 18) [169, 175], which variation is linear in the reduced stress (stress divided by absolute temperature). De Gennes predicted

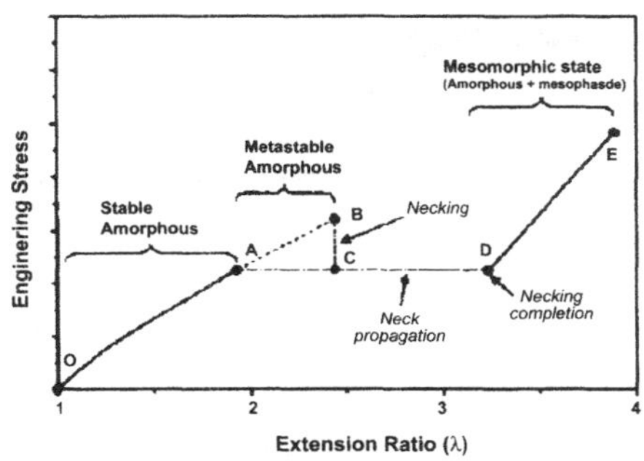

Fig. 16 Schematic diagram of engineering stress vs extension ratio for an initially amorphous PDES elastomer (after Hedden et al. [82])

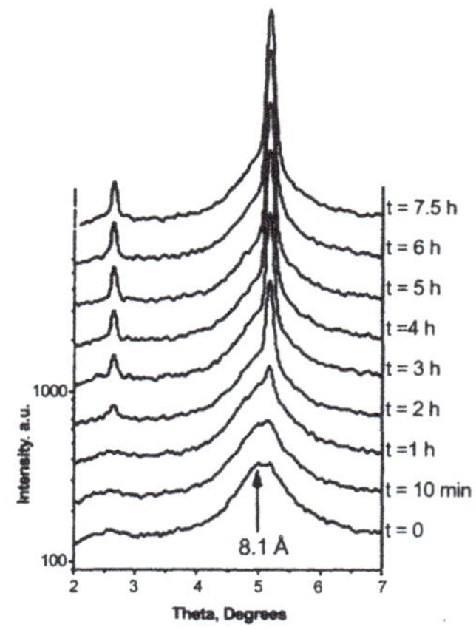

Fig. 17 Kinetic WAXS study of the stretch-induced isotropic-mesophase transition in cross-linked PDES (after Koerner et al. [174])

such a linearity [177] for liquid-crystalline elastomers on the basis of nematic interactions between the mesogens. Deviations of conventional elastomers away from classical theories have often been interpreted using this notion. In particular Warner et al. [178] explained the linearity between reduced stress and transition temperature combining nematic interactions between segments with the concept of a semi-rigid wormlike chain.

Measurements in solution indicate that PDES is a flexible chain polymer with a characteristic ratio $C_\infty = 7.7$ [179], however, in the melt its chain tends to extend due to a negative thermomechanical coefficient. Calorimetric measurements of heat and energy effects from stretching PDES networks show a decrease of the internal energy ΔU with extension [169]. Even at low extension or high temperatures, for which any mesophase is excluded,

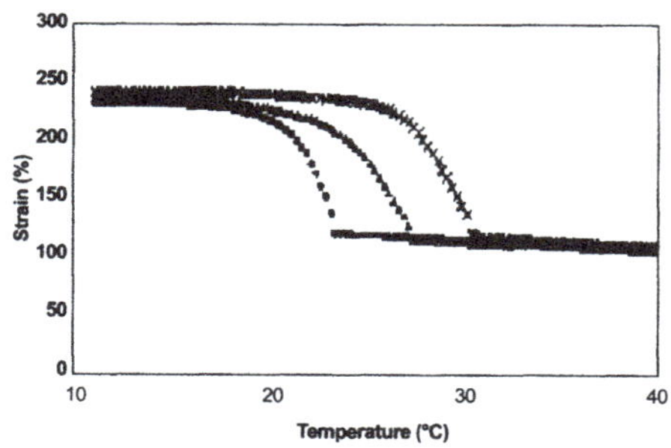

Fig. 18 Strain-temperature curves for cross-linked PDES at three different values of the uniaxial force: 2.1□ × 10^5 Pa; Δ2.2 × 10^5 Pa; ×2.3 × 10^5 Pa. The stress values are based on the original cross section of the unstretched amorphous sample (after Hedden et al. [175])

the ratio between internal energy and work is close to – 0.25. This value corresponds to a large negative temperature coefficient of the unperturbed chain dimension: $d(\ln\langle r^2\rangle_0)/dT = -0.8 \times 10^{-3}\ \mathrm{deg}^{-1}$. This means that PDES chains energetically prefer to extend, which is one reason for the formation of the mesophase. Calculation of the conformational energy shows a maximum mobility for the ethyl side groups for an increased number of straight conformations of the main chain [180]. Specific sequences of ordered conformations survive in the mesophases and effectively play the role of mesogens. However, these 'mesogens' are neither permanent nor make up the total chain. In this picture the mesophases can also be considered as lyotropic liquid crystals in which the same molecule provides mesogenic parts and 'solvent' parts in a microphase-separated structure.

Hexagonal packed condis crystals (or columnar phases) widely exist in polymers. Three types of polymers form a hexagonal phase [46]. (i) Flexible linear macromolecules: PTFE and related halogenated polyethylenes, PB, 1,4-cis-poly(2-methylbutadiene), PE, poly(p-xylene), etc. (ii) Flexible branched (comb-like) macromolecules: alkyl-branched polysiloxane, polysilanes and polygermanes, aryl- and some alkyl-branched polyphosphasenes, etc. (iii) Rigid macromolecules with flexible side chains: n-alkyl substituted poly(L-glutamate) and cellulose. At a giving temperature and pressure the hexagonal phase can be thermodynamically stable. Analogous to the situation discussed for PDES, stretch-induced mesophases are expected in all these polymers. In constrained ultra-drawn PE fibres the hexagonal phase persists for a rather long time and can even recover after melting and subsequently cooled down to the crystallization temperature [181].

2.4 Summary and discussion

Stimulated by an external uniaxial flow field, mesophases have been observed in three types of polymers with different characteristics. The semi-rigid chain polymer PET contains intrinsic rigid block elements; nevertheless in general it does not show liquid crystallinity. However, promoted by alignment and stretching, orientational ordering of the molecular chains can take place. At temperatures below the glass transition T_g, stretching can induce a nematic and a smectic phase. Slightly above T_g these mesophases can still appear provided the draw rate is large. In the smectic phase of PET essentially the intrinsic length scale of the monomer unit determines the smectic layer spacing. Slight deviations that have been observed can be attributed to conformational disorder or tilt of the molecular chains.

In the flexible chain polymer iPP, a shear-induced smectic phase with a periodicity of about 4 nm has been observed at temperatures below as well as above the melting point. Increasing the average molecular weight leads to a slightly larger periodicity. The smectic layers assemble in a fibrillar morphology with a length and width up to 200 μm and 10 μm, respectively. These fibrillar entities have been modelled as anisotropic drops of smectic domains in which the smectic layers can rotate. On the basis of this anisotropic drop model the relative orientation of the smectic layers, the crystalline lamellae and the long axis of the droplet, as well as the reversibility of the smectic periodicity during cooling and heating, can be understood.

A molecular chain of PDES contains no mesogenic elements; still two columnar mesophases μ_1 and μ_2 exist above the melting point of the crystalline state. In cross-linked PDES elastomers stretching can induce an amorphous-mesophase transition. With increasing stress the transition shifts to higher temperature. Kinetically this transition follows a 2D growth model.

The formation of the three types of mesophases as discussed above has a common origin: induced rigidity. PET molecular chains already possess a certain rigidity without an external field, which is slightly lower than needed to reach an isotropic-nematic transition. Imposing a flow field at temperatures close to T_g, conformational ordering can easily enhance the rigidity to meet the isotropic-mesophase transition. iPP chains are in principle truly flexible, but in the melt conformationally ordered helical segments with up to 5 monomers already exis. These helical sequences are too short to induced orientational ordering. To reach a LC phase transition the spontaneous orientational and conformational ordering must be enhanced by a flow field. In the columnar mesophases of PDES the molecular chains comprise conformationally ordered and disordered segments, which results in a lyotropic-type LC phase [180]. Evidently the formation of mesophases from these semirigid and flexible chains is based on conformational ordering enhancing their rigidity. In all cases the ordering of the conformers of the chain is only partial, which determines the nature of the mesophase ordering.

The origin of the PDES columnar phases and the iPP smectic phase is essentially the same: A mixture of conformationally ordered and disordered segments. The transient phase from the iPP smectic phase to crystal formed during a fast quench is similar to the columnar phase in PDES. A smectic phase could be expected in PDES under a flow field at high temperatures. In fact a flow-induced smectic phase has been suggested for PDMS [83], which compound closely resembles PDES. Remaining questions are after the nature of the layering and the origin of the periodicity. Strong fluctuation usually destroys any long-range ordering that is based on secondary weak interactions. In general a smectic layer spacing is directly related to a molecular dimension in the constituent system, like the monomer unit in PET or the chain length in block copolymers [44]. For strong interactions like a first-order phase transition resulting in crystallization, the lamellar thickness is the determining dimension [182]. In iPP smectic layers, the interactions among molecular chains are rather weak and are not expected to stabilize a layer structure with linear chains. One possible reason for its stability is the connectivity among layers by the same chains. This provides a damping of fluctuation analogous to the situation in a cross-linked side-chain smectic polymer [183]. Note that flow-induced mesophase ordering results in a biphasic (mesophase and liquid) coexistence, which is due to the non-thermodynamic nature of the behaviour of polymers and to their complexity such as polydispersity.

The possibility of formation of mesophases is strongly influenced by molecular parameters like the length of the molecular chain. The precise nature of these effects is still an open question. The molecular weight may directly control the size and the thermal stability of the mesophases in PDES. In both PET and iPP the organization of the molten chains and the related rheological properties – which depend on molecular weight – influence the formation of the mesophase under flow conditions. We anticipate the thermal stability of these mesophases to be influenced by the number of layers connected by the same chains. In that case a longer chain length gives larger domains of correlated layers and an increased stability of the mesophase. As shown theoretically by Brochard and de Gennes [67], the polydispersity of the molecular weight also plays an important role in mesophase ordering. For PET and PDES no information concerning this point is available. For iPP a larger polydispersity gives at the same experimental conditions a larger concentration of smectic bundles [6]. A quantitative elucidation of these effects requires more investigations concerning the influence of molecular parameters (molecular weight and its distribution, tacticity and its distribution, etc.).

We have analysed the structure of flow-induced precursors of crystallisation on the basis of real-time characterization techniques such as X-ray scattering. They can be identified unambiguously as mesophases with ordering intermediate between amorphous and crystal. Though essentially we only reviewed three polymers, in analogy we expect a great number of polymers to show similar behaviour. Of course, the conditions under which semicrystalline polymers show mesophase behaviour may be very different. In many cases flow-induced mesophases are metastable transient states with a short lifetime. Consequently the time resolution of the characterization techniques is of ultimate importance. Due to the increasing accessibility of synchrotron facilities, we anticipate in the coming years major achievements in this field from X-ray methods.

3
Influence of mesophase behaviour on crystallization

In flow-induced crystallization of polymers, the occurrence of an intermediate mesophase before the onset of the crystallization itself has a profound influence on the process, which we shall consider in this section. As mentioned above, due to the non-thermodynamic nature of the processes involved we encounter a biphasic coexistence of a partially ordered mesophase and the polymer melt. Evidently this provides a common starting point to discuss flow-induced crystallisation of the three different polymer systems. Two aspects can be distinguished in the role of the mesophase. The first situation is called 'mesophase nucleating'. The mesophase serves as nucleation site for crystallization, even if the mesophase itself may not be easy to transform into crystals. The driving force for a transition from mesophase to crystalline is expected to be smaller than for the transition from an amorphous state to crystalline because of the mesophase ordering corresponding to a local minimum in the free energy landscape. Moreover, in some case confinement imposed by different length scales may also lead to a high energy-barrier for the transition from mesophase to crystal. In addition to serving as nucleation site, the mesophases act directly as a template or precursor facilitating its subsequent crystallization. Because the mesophase already has a certain ordering (orientational, conformational, density) kinetically crystallization from an ordered mesophase can be relatively fast compared to crystallization from the disordered amorphous state. We refer this phenomenon as 'mesophase templating'.

3.1 Crystallization on mesophase nuclei

In the case of iPP, the nucleation effect of smectic phase on the subsequent crystallization is quite pronounced, more than any templating, as can be demonstrated as follows. First the melt is sheared well above the crystallization temperature inducing smectic layering, and subsequently the sample is cooled down to the crystallization temperatures. The crystallization kinetics of iPP with and without shear-induced smectic bundles has been studied with in-situ SAXS [6]. The relative SAXS intensities are plotted in Fig. 19 vs crystallization time. The sample with shear-induced smectic bundles crystallizes faster than the other one. The nucleation effect has been confirmed by optical microscopy. In Fig. 20 a series of POM images is presented of an iPP sample with shear-induced smectic bundles during crystallization at 160 °C. The original smectic layers assemble into bundles with a diameter and length up to 10 μm and 200 μm, respectively. After about 100 min crystals grow from the surface of these bundles; at later crystallization times so-called transcrystallites develop (shish-kebab structure). Alternatively, a shear flow can be imposed on a supercooled melt at the crystallization temperature. Now the shear-induced smectic ordering and crystallization take place in sequence. The intensity of the smectic peak can be taken as a measure of the concentration of smectic bundles. We find that larger smectic peak intensity corresponds to faster crystallization rate [151]. These experimental results support a nucleation effect of the smectic bundles on the subsequent crystallization.

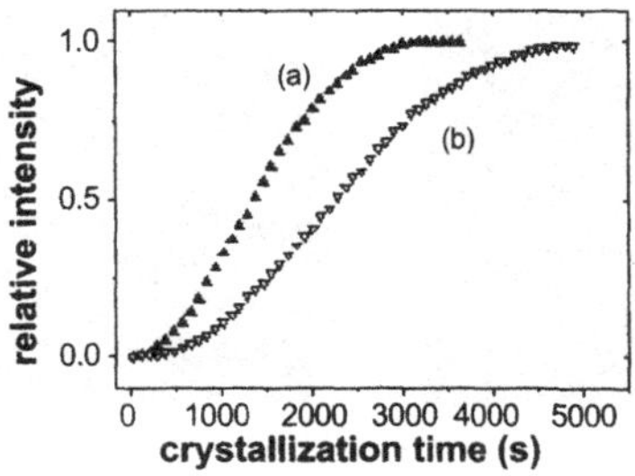

Fig. 19 Relative SAXS intensities vs crystallization time of iPP (**a**) in presence of shear-induced smectic bundles (**b**) without smectic bundles (after Li and de Jeu [6])

As shown in Fig. 7, in the melt the smectic layers can rotate and align in the direction of the velocity gradient. This provides a possibility to tune the relative orientation of the smectic layers and the crystalline lamellae. Figure 21a shows a two-dimensional SAXS pattern of a crystallized iPP sample with the smectic layers aligned along the velocity gradient. A similar pattern of iPP crystallized after a shear pulse at 140 °C (alignment parallel to the shear) is displayed in Fig. 21b. The corresponding azimuthal intensity distributions of both the smectic peak and the crystalline peak are plotted in Fig. 21c and d, respectively. As we see, the crystalline lamellae can be either parallel (Fig. 21b and d) or perpendicular (Fig. 21a and c) to the smectic layers. However, the normal of the crystalline lamellae always follows the direction of the bundle axis, consistent with the transcrystallite model. Figure 21 gives two extreme possibilities; intermediate situations can result from different shear histories.

To understand the mechanism leading to enhancement of crystallization under shear, the structure of the so-call shish or row nuclei, the primary nucleus induced by flow, is essential. The common textbook picture is that the shish consist of extended-chain crystals that may be defective [184–189]. It is generally accepted that structures with some degree of order are induced by the flow before the onset of crystallization, which can be long living at shallow supercooling [127–150]. These shear-induced precursors have been described

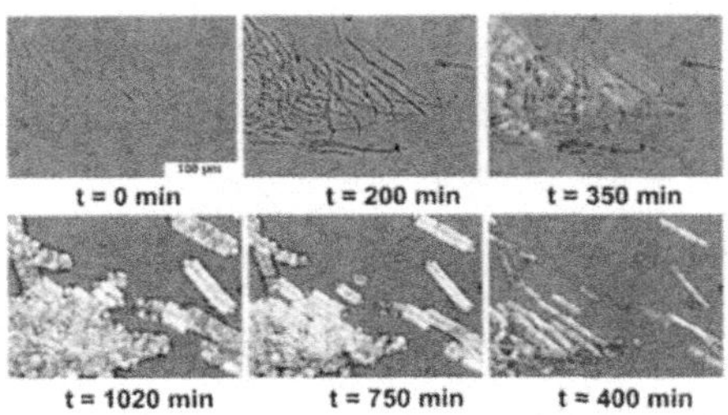

Fig. 20 POM pictures showing smectic-bundle-induced crystallization at 160 °C. The scale bar of the second row is twice that of the first row (after Li and de Jeu [6])

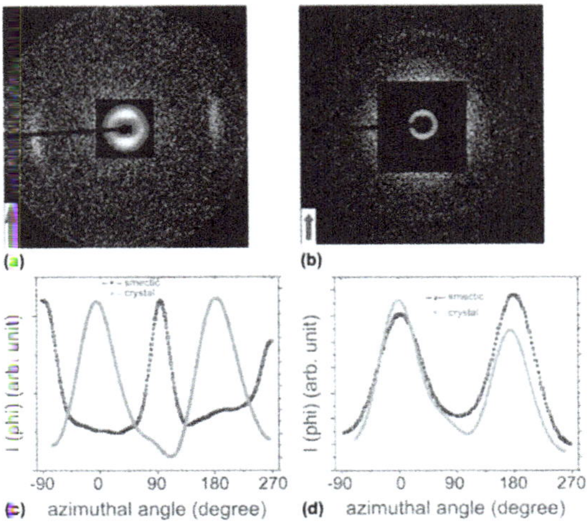

Fig. 21 Two-dimensional SAXS pattern of (**a**) iPP crystallized at 140 °C after shear-alignment in the melt (**b**) step-sheared (rate of 30 s^{-1} and strain of 1500%) and crystallized at 140 °C. The corresponding azimuthal intensity dustribution of the smectic and crystalline peaks are shown in (**c**) and (**d**) , respectively. The arrow indicates the flow direction (after Li and de Jeu [6])

as bundles with stretched chains or are supposed to be due to a liquid-liquid phase separation. Our studies on iPP indicate that these primary row nuclei are assemblies of smectic layers [6, 152, 153]. The fibrillar morphology, high temperature stability, and molecular weight dependence of the smectic domains, all not only meet the general description of these precursors but also account for the final shish-kebab structure. The relation between

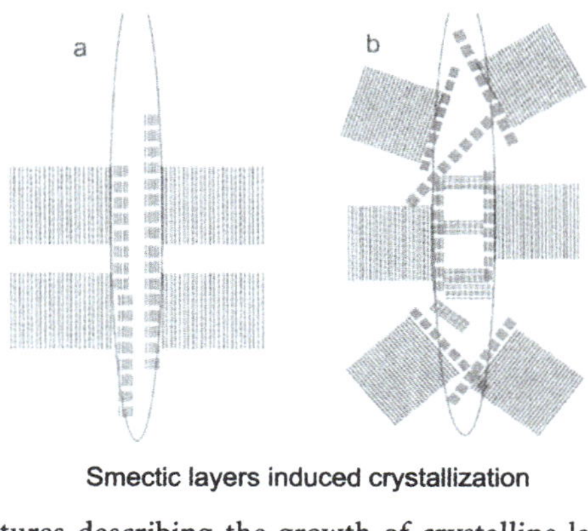

Fig. 22 Schematic pictures describing the growth of crystalline lamellae on the smectic bundles (**a**) without and with shear-alignment (**b**) with shear-alignment (after Li and de Jeu [6])

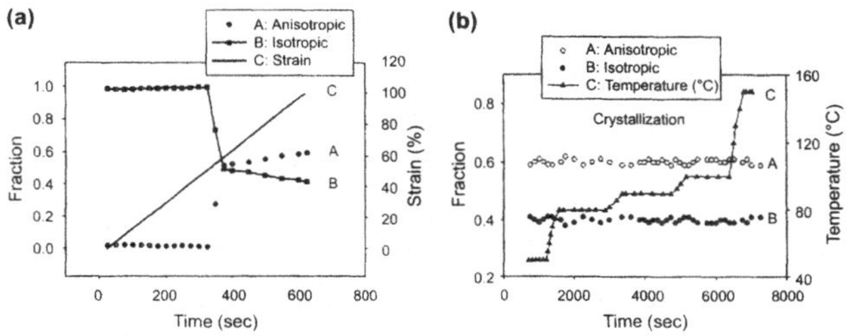

Fig. 23 Variation of the isotropic and anisotropic mass fractions of PET during (**a**) the deformation stage and (**b**) the crystallization stage (at constant strain) (after Ran et al. [96])

average molecular weight (its distribution) and the smectic periodicity (the concentration of smectic bundles) provides a first step to a molecular understanding of shear-induced crystallization. The average molecular weight determines the stability of the shear-induced primary nuclei – the smectic bundles, while the distribution of molecular weight (or the ratio between long and short chains) controls the density of nuclei. Figure 23 presents a new shish-kebab or transcrystallite model, in which the row nuclei are anisotropic smectic droplets rather than extended-chain crystals. In the case of a short shear pulse, smectic layers are induced by the flow and acquire a weak initial alignment along the shear direction. The crystalline lamellae grow in the same orientation as the smectic layers (Fig. 22a). After imposing the shear field for a longer time, the smectic layers align along the direction of the velocity gradient (see Fig. 7). In most experimental situations the formation and alignment of the smectic layers occur simultaneously. Depending on the shear field the smectic layers may to some extent loose their initial orientation (Fig. 22 b) and the growth of the crystals deviates from the direction perpendicular to the axis of the smectic bundles.

3.2
Crystallization from a mesophase precursor

In PET the mesophase plays the role of template for the crystallization. Figure 23 shows the variations of isotropic and anisotropic mass fractions during (a) deformation and (b) crystallization [96]. In Fig. 23a, the structure of the initial sample is fully isotropic. As the neck region moves into the detection area ($t = 325$ s, the isotropic fraction decreases dramatically and the anisotropic fraction becomes rapidly larger over the entire pattern until the extension reaches 58% at $t = 375$ s With a further increase in strain, the anisotropic fraction still increases somewhat indicating that more chains are oriented. This is consistent with the observed increase of the (001′) peak intensity at $d = 1.03$ nm. In the crystallization stage both the anisotropic and isotropic fractions stay about constant (Fig. 23b). As the amorphous fraction remains constant during crystallization stage we must conclude that it is not directly converted into crystals. This provides strong evidence that strain-induced crystallization occurs mainly in the mesophase region developed upon stretching. It is easier for PET molecules in the mesophase to crystallize than for molecules in the amorphous phase, since in the former case the chains already posses some arrangement and orientation similar as in the crystal. As the crystals develop from oriented chains in the mesophase, it acts as precursor for strain-induced crystallization and can greatly accelerate the crystallization process. In fact, in the oriented sample no additional crystallization directly from the amorphous phase has been observed. This conclusion was first suggested by Shimizu et al. [190] and has recently been confirmed by Windle et al. [110] and Wu et al. [191]

Structural differences in the initial mesophase affect the crystallization behaviour. Drawing PET at a slow rate at 90 °C does not lead to clear indications of a smectic phase; probably a nematic phase is obtained. As indicated by the scattering intensity of the (110) and (010) peaks [95–106], during subsequent crystallization the lateral packing develops faster than the packing along the chain direction. In contrast, the layer structure of a smectic phase leads to an easy crystalline arrangement along the chain axis, as derived from the intensity of the (– 103) peak. The influence of the mesophases on crystallization varies with temperature. The crystallization in oriented PET, hence with mesophases, has been studied at temperatures around the glass transition [95–106]. In this temperature range the crystallization kinetics of un-oriented molecular chains is extremely slow and may not be observable at the time scale investigated. Hence transitions from mesophase to crystal are observed rather than crystallization from the unoriented amorphous region. For crystallization temperatures around 170 °C or higher, where PET shows the fastest crystallization rate, crystallization is expected both in the oriented mesophase and in the amorphous region [192]. In this regime the nucleation effect of the oriented mesophase should be more pronounced.

PDES is another typical case of mesophase templating in which the crystallization behaviour is dramatically changed by the presence of mesophases. Following a fast cooling directly from the melt, the γ modification is obtained rather than α and β crystals. The latter two form after cooling from the mesophase [163, 164]. As illustrated in Fig. 14, the formation of either β_1 or β_2 crystals is due to a starting point in a different mesophase. For crystallizing after fast quenching from an amorphous state, the kinetics of PDES is similar as found in standard polymer crystallization experiments. This process results in a crystallinity of about 30–40% [193]. In contrast, if PDES first forms a mesophase, the crystallization rate is very high and prevents the formation of an amorphous glass at fast quenching. Crystallization from the mesophases generates in PDES typically a crystallinity exceeding 90%. Similar results have been obtained in PE when crystallized from its mesophase counterpart [78]. In these examples the template effect of the mesophase on crystallization is evident [171]. In a cross-linked PDES network a high extension of molecular chains can be reached and fast crystallization is obtained through the stretch-induced mesophase. However, the final crystallinity is lower than that in a non-cross-linked sample because of the topological constraints from the cross-links. Other polymer networks show similar crystallization behaviour; for example, fast crystallization has been observed in natural and synthetic rubbers [194–197].

3.3 Summary and discussion

On the basis of occurrence of flow-induced mesophases a better understanding of flow-induced crystallization of polymer has been achieved. From this starting point many observations can be understood. Let us reconsider the phenomenological observations on flow-induced crystallization as described in the introduction. First, the acceleration of the crystallization kinetics induced by flow can be attributed to the nucleating and templating effects of the mesophases. Second, the molecular weight and its distribution influence the stability and the concentration of the mesophases, and consequently control the density of nuclei. So far flow-induced mesophases have not been taken into account in theoretical studies [198, 199] and computer simulations [200, 201] of polymer crystallization. The influence of orientation on nucleation can be attributed to an entropy difference between the two molten states ($\Delta S = S_{\text{oriented}} - S_{\text{unoriented}}$), in which the oriented entropy is lower than the unoriented (quiescent) one. An expression for $N^{\text{f}}/N^{\text{q}}$, where N^{f} and N^{q} are the nucleation rates under flow and at quiescent conditions, respectively, has been given by Yeh and Hong [198] for high temperatures $T > (T_{\text{m}} + T_{\text{g}})/2$. It reads

$$\frac{N^{\text{f}}}{N^{\text{q}}} = \exp\left[\frac{\gamma_{\perp}\gamma_{\parallel}^{2}}{k_{\text{B}}T}\left(\frac{T_{\text{m}}^{\text{o}}}{\Delta H^{2}\Delta T^{2}} - \left(\frac{\Delta H \Delta T}{T_{\text{m}}^{o}} + T_{\text{m}}^{0}\Delta S\right)^{-2}\right)\right], \tag{11}$$

in which $\gamma_{\perp}$ and $\gamma_{\parallel}$ are the free energies of folded-chain and lateral surfaces, respectively, T_{m}^{0} is the equilibrium melting temperature of crystals and $\Delta T = T_{\text{m}}^{0} - T$ is the supercooling. Since the first term in the exponent is greater than or equal to the second one, faster nucleation is expected under flow. According to Eq. 11 the decrease of entropy ΔS due to flow alignment and stretching of chains is the essential contribution to the enhancement of the nucleation density and the role of the enthalpy can be disregarded. However, if the nuclei induced by flow are due to mesophase formation, the enthalpy effect can be much more pronounced. As mentioned in Sect. 1.2, the enthalpy of a mesophase is usually an order of magnitude less than for a crystal. As a result it is much easier to induce by an external flow field mesophase ordering than crystallization. At usual experimental temperatures heterogeneous nucleation dominates the polymer crystallization. Homogeneous nucleation occurs at a supercooling 60–100 °C larger than employed in the shear-induced crystallization experiments of iPP. The loss of entropy induced by the flow can be equivalentlyconsidered as a virtual increase of the melting temperature and thus of the supercooling. For shear rates of tens s^{-1} it is impossible to obtain a virtual increase of the melting temperature of 60–100 °C. Evidently, the loss of entropy alone is not enough to account for an increase of the nucleation density by orders of magnitude. By incorporating the enthalpy contribution associated with a flow-induced mesophase, this discrepancy can in principle be solved. A more quantitative estimate is complicated and will vary for different polymers.

We conclude that mesophase formation can influence subsequent crystallization in various ways. (i) Both nucleating and templating effects can accelerate the crystallization kinetics. The nucleation effect of a mesophase is widely accepted, the templating effect has received less attention so far. (ii) The presence of a mesophase can influence the crystal structure. In addition to the structures obtained at quiescent conditions, for many crystalline polymers an external field induces one or more crystalline modifications [202]. The formation of such a modification can again be attributed to nucleation and/or templating effects of the mesophases. Examples are provided by the relations between the μ mesophase and the β crystal modification in PDES and between the smectic phase and β crystals in iPP [6, 172]. (iii) The presence of a mesophase before crystallization can result in a different crystal morphology. Because the molecular chains in a mesophase have in general a high mobility, thick or even extended-chain crystals can result upon crystallizing from such a situation [78, 164, 172]. If a mesophase is pre-aligned by a flow field, the nucleation and templating effects can also guide the growth direction of the crystals [6].

4 The structure of the disordered polymer state

The presence of a mesophase prior to flow-induced crystallization provides clear explanations of several experimental observations and gives possible answers to some fundamental questions regarding polymer crystallization. However, questions remain around the emergence of a mesophase from the amorphous (glassy) phase as well as from the melt. Which molecular factors determine what type of phase can be induced under certain circumstances? The answer is not only important for understanding polymer crystallization but also for other ordering processes in polymers. This brings us to the disordered counterpart that is the starting point for the formation of any ordered mesophase or crystal. What is the structure of the disordered polymer melt and of the amorphous state? Evidently different starting points can lead to different views on the subsequent ordering process. It is widely accepted that the polymer chains in the melt and amorphous state are in a random-coil (RC) state, which is equivalent to the so-called theta condition [203]. In this situation, a random-walk model gives a scaling law between the radius of gyration R_G and the molecular weight $M : R_G \propto M^{1/2}$. Following the RC model an external flow field can orient and stretch the molecular chains in the flow direction provided the flow rate is fast enough to overcome the relaxation process. Figure 24a shows a schematic picture of such a flow-induced alignment of molecular chains. This picture is widely employed to interpret the entropy loss of the molecular chains and is the basis of extended-chain model of row nuclei [4, 5, 127–150].

Is the random coil picture fully applicable to a real polymer melt used as starting point for polymer ordering? During past half a century Yeh and others have questioned this point [204–207] (see also volume 68 of the Faraday Discussion in 1979). On the basis of ball-like structures observed in amorphous PET and related polymers, Yeh et al. proposed a folded-chain fringed-micellar grain (FFMG) model for the amorphous state. It comprises grains with parallel chains of a diameter of 2–10 nm and intergrain regions containing segments in a truly random coil state [204, 205]. This model has been seriously questions by, amongst others, Flory [208] and Fischer et al. [209]. Nevertheless let us test its possibilities as starting point for flow-induced smectic ordering in iPP as illustrated in Fig. 24b. For the FFMG model the flow field induces alignment and stretching of molecular chains similar as in the case of random coils. However, in addition the folded-chain micellar grains will be aligned. Because the latter is a co-operative process, it will already be pronounced for weak flow fields and this picture fits well to the experimental observations of flow-induced smectic ordering in iPP [6, 151, 152]. Yeh et al. applied the model to explain their observations on stretch-induced crystallization of PET [31, 32]. Evidently it is most appropriate for situations in which local correlations are important. These

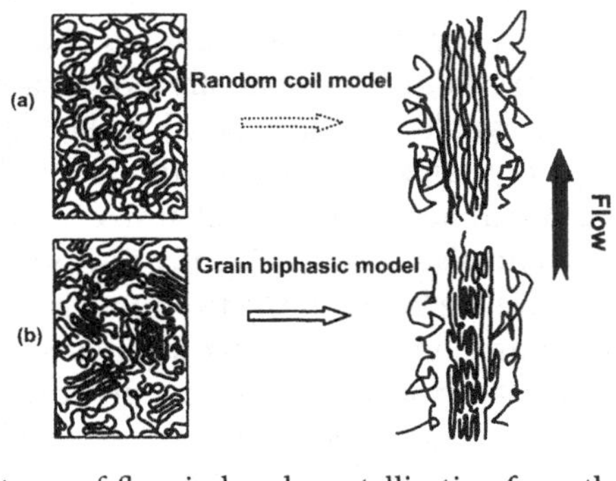

Fig. 24 Schematic pictures of flow-induced crystallization from the polymer melt (**a**) for the random coil model (**b**) using the folded-chain fringed-micellar grain model

observations inspired us the compare systematically the experimental evidence supporting the RC and the FFMG model in the present context of flow-induced crystallisation.

The RC model is essentially based on theoretical statistical physics [202]. It describes an averaged polymer melt and amorphous state and does not take local correlations into account. The evidence supporting this model as provided especially by Flory [208, 210] can be listed as follows. (i) The observation of a retraction force f in stretched elastomers. (ii) The force-temperature coefficient and the correspondence of $[-\partial \ln(f/T)/\partial T]_{V,L}$ to $\mathrm{d}\ln\langle r^2\rangle_0/\mathrm{d}T$ found for linear polymer in dilute solution. (iii) A comparison of the experimental cyclisation constants K_x for siloxane, both in absence and in presence of an inert diluent, with values calculated from the dimension $\langle r^2\rangle_0$ for linear polymers. (iv) The thermal activities of solutions. (v) The description of the overall molecular dimensions of bulk polymers. However, the underlying experiments are not particularly sensitive to the presence of any local (microscopic) order. Small-angle neutron scattering (SANS) indicates that the radius of gyration R_G of molecular chains is proportional to $M^{1/2}$ both at theta conditions and in the melt. Though this fits well to the predictions of the RC model, we note that R_G at theta conditions varies with solvent [211]. This indicates that the generally accepted use of such data as reference for the melt is not straightforward. In most cases R_G in the melt is larger than that at theta conditions, indicating that the organization of the molecular chains in the melt deviates from the RC model [203]. PDES chains that are flexible in solution and rigid in the melt [171, 172, 179] provide an extreme example. Flory's prediction of unperturbed random coils in bulk amorphous polymers is based on the assumption that the interactions between chain segments belonging to different molecules are independent of the local chain conformations. In many cases it may be necessary to consider also conformation-dependent intermolecular (attractive) interactions [212].

Measurements of R_G by SANS do in fact not provide microscopic information. Ballard et al. determined the overall molecular dimensions of iPP from 23 to 220 °C. Most surprisingly, within the experimental errors no difference was found between crystallized and molten chains [213, 214], while obviously the chains in crystalline lamellae are not in a random coil state. These results allow two possible conclusions. (i) If the experimental data of chain dimensions in the crystallized state can still be fitted to the RC model, then local ordering has no influence on the global size of the molecular chains and the FFMG model would also apply. Evidently SANS results on molecular dimension cannot discriminate between the two models. (ii) The experimental similarity of the organisation of molecular segments in the melt and the crystallized state can be used in a reversed way. It indicates that the local organisation in the melt is the same as in the crystal, which agrees with the FFMG model. The difference lies evidently in the ordering on a larger scale. Taking the original data from Ballard et al. [213] we plot in Fig. 25 R_G in quenched iPP vs $M^{1/2}$. As stated by the authors, within the experimental errors these values are the same as in the melt. To prevent misguiding, in Fig. 25 the linear fit is omitted. The interesting point is not only the non-linear behaviour, but also the larger deviations from linearity at low molecular weight. This behaviour can be understood in the framework of the FFMG model. Chains with a large M have relatively large overall dimensions that are not significantly influenced by the presence of grains of diameter 2–5 nm with parallel segments. But for the smaller molecular size of low-M polymers, the presence of such regions causes R_G to deviate from the Gaussian value as shown. If for long chains the crystal thickness is large, deviations will also occur, exactly as observed by Ballard et al. [213, 214].

Direct evidence of the FFMG model is provided by the observation of granular domains in a wide range of amorphous polymers [31, 32, 204, 205]. Yeh et al. used radial distribution function analysis of WAXS data from amorphous polymers [215, 216], which is more sensitive to local ordering and revealed local parallel alignment of chains. Granular domains have also been observed in rubbers. Inhomogeneous density distributions in cross-linked rubbers and other amorphous polymers with a polydisperse molecular weight have been well established on the basis of SANS. Lozengic and so-called 'butterfly' scattering patterns have been observed in these systems after stretching and long-time relaxation [217–221]. However, three small modifications of the FFMG model are necessary to obtain agreement with the experimental observations presented here. First, in the melt the boundaries between grains and intergrain regions should be diffuse and dynamic. The sharp boundaries observed by Yeh et al. [31, 32, 204, 205] may develop during cooling from melt. Second, any density difference between these two regions should be beyond the detection limits, even though a parallel packing of molecular segments is adopted in the grains. Third, the grain size in the melt should be rather small such that the melt phase remains effectively homogeneous. These modifications assure that the grains alone are not a thermodynamic phase. Segments with different conformations can coex-

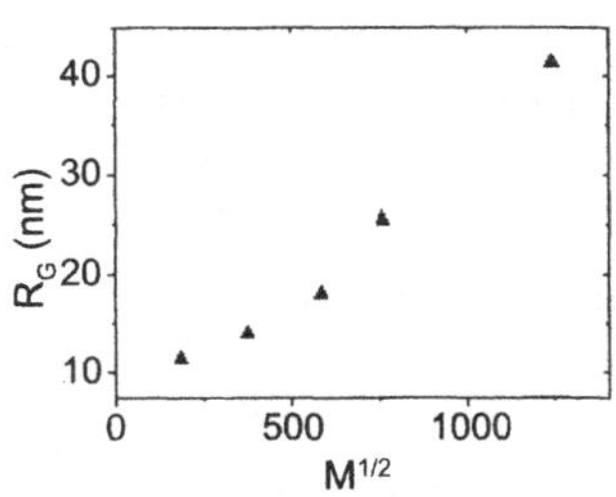

Fig. 25 Radius of gyration vs $M^{1/2}$ for iPP according to the original data from Ballard et al. [213]

ist as a thermodynamic equilibrium state in the same molecular chains, like in the columnar phase of PDES and the smectic phase of iPP [6, 180]. The assumptions of a small size of the grains and a negligible density contrast are needed to accommodate that the grain structure is not directly observed by depolarized light scattering and SAXS [209]. Lowering the temperature may induce growth of the grains and lead to ordering on a larger scale. At temperatures near T_g, the correlation length of the fluctuation increases, as revealed by mode-coupling theory [222]. The nucleation density of samples quenched down to the glass state and then heated up to the crystallization temperature is generally much higher than that in samples directly cooled from the melt. A flow field can induce alignment of these grains and induce ordering, accompanied by an increase of the grain size. The helical sequences in the iPP melt consist of up to 5 monomers, while longer helices are obtained after imposing shear. Mcalea et al. investigated PET chains in the amorphous state by SANS [223]. Though the overall dimension of the molecular chains could be fitted with the RC model, the scattering in the intermediate q-range did not follow the RC expectations. These types of deviation were more pronounced in semi-crystalline samples and larger deviations were found in PET with low M, as expected in the FFMG model.

The flow-induced smectic ordering in iPP evidently favours the folded-chain fringed micellar model for the iPP melt [6, 152, 153]. The main effect of a weak shear field is to align the grains, which simultaneously promotes close packing and conformational ordering. More generally, during flow-induced crystallization, a weak flow field can induce a large increase of the crystallization rate. This is not the direct expectation of the RC model and polymer dynamics: the effect of a flow field can be erased by relaxation of the chains if the flow rate is smaller than the time scale of the relaxation. However, a weak flow field can be enough to align the grains and induce primary nuclei for polymer crystallization. Further increasing the stress, the structure of Fig. 24b switches to that of Fig. 24a. Interestingly, during this process, the number of nuclei does not change, assuming no new nuclei to form other regions. This gives a possible answer to the saturation of the flow-induced acceleration of the crystallization rate upon increasing the flow rate [129–152].

The lozengic and butterfly scattering patterns can also be interpreted in terms of the FFMG model. Imposing a strong stretch or shear field upon a statistical cross-linked network or a normal crystallizable polymer, the scattering intensity of SANS [217–221] or SAXS [22, 30] in the equatorial direction (perpendicular to flow direction) is larger than that in meridionial direction. Two streaks are observed along the equator (see Fig. 11). After relaxation or if the initial flow field is weak, a butterfly scattering pattern can be obtained in SANS, but with a larger scattering intensity along the meridian than equatorially. Similar patterns are also observed in crystallizable polymers before the onset of crystallization. Though details of the scattering patterns and the underlying structures may be different, the crucial feature is an inhomogeneous density distribution pointing to the existence of domains with different dimensions. The approach of Brochard and de Gennes [67], which uses the RC model, predicts always a larger scattered intensity perpendicular to the flow. With the FFMG model as starting point the schematic picture of Fig. 24b can explain the observations in a natural way both for strong and weak external flow. The size of the domains is expected to vary for different polymer systems and to depend also on the external field.

The FFMG model can in principle be extended to polymer crystallization at quiescent conditions. The debate about pre-ordering before crystallization is mainly related to the nature of the primary nuclei. The well-established folded-chain crystal model of polymers has pushed the polymer community to accept a folded chain as the primary nucleus. Zachmann [224] calculated the effective free energy of folded-chain and fringed-micelle (or bundle nucleus) models [225–228] and found that the folded-chain primary nucleus is favoured. However, in the case of a mesophase precursor, in Zachmann's approach the surface free energy of the bundle nucleus is greatly overestimated [224]. In the smectic phase of iPP or other mesophases the excess interfacial energy, caused by entropy constraints on the connection to the nuclei of liquid segments, is much lower than in a crystalline phase. Hence for nucleation through a mesophase the energy barrier is low, favouring fringed-micelle nuclei. Then the scheme of Fig. 24b can also be applied to the nucleation process at quiescent conditions. This picture of bundle or row nuclei is consistent with Muthukumar's computer simulations [229, 230]. In the polymer melt, the 'baby nuclei' do not necessarily reach the final lamellar size by thickening alone. When adjacent 'baby nuclei' aggregate or align and reach the critical size of a nucleus, crystal growth can take place.

Finally the FFMG description of the polymer melt leads to a different view of the growth front of lamellar crystals. On the folded-chain surface of crystals at least 50% of the area consists of tight folds, even at large supercooling or fast quenching. The original picture of the Lauritzen-Hoffmann theory assumes that the chains shuttle forth and back between two folded-chain surfaces and are reeled in at the growth front of the lamellar crystal [12]. This is how tight folds are supposed to be built. These ideas have lead to a somewhat confusing debate about the too limited speed of the reeling process and possible ways to repair this problem [12, 231, 232]. Hence an alternative process to create the folded-chain surface would be very welcome. Taking the FFMG model as starting point of crystallization, the folded-chain grains can be adsorbed entirely on the growth surface (see Fig. 26a). Subsequently sliding diffusion on the surface allows to re-arranges the chain segments (see Fig. 26b). The process is fast enough to build tight folds – even following a fast quench – because the folds are already present before the crystallization starts. The latter process indeed has been observed with computer simulations [230, 233].

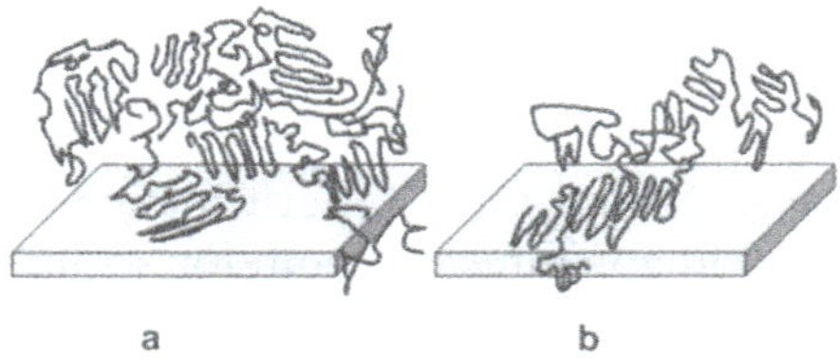

Fig. 26 Schematic picture of the growth front of a polymer crystal according to the folded-chain fringed-micellar grain model of the melt

5 Concluding remarks

In this review we have discussed flow-induced mesophase ordering and its influence on subsequent crystallization. Three typical semi-crystalline polymers without intrinsic rigid mesogens have been considered that all present mesophase behaviour under external flow. The long-chain nature of polymers provides the basic requirement for mesophase formation with orientational ordering. We speculate that upon imposing a proper external flow field, mesophase ordering is quite general and not restricted to these examples. The origin of mesophase ordering in semi-rigid and flexible chains is attributed to induced rigidity by conformational ordering. The mesophases have an enabling effect on subsequent crystallization. By promoting nucleation and templating, a mesophase may accelerate the crystallization rate, lead to different crystal modifications, change the morphology, and guide the orientation of the crystals.

Searching for the mechanisms of mesophase formation, we have described why the answers may lie in the disordered states. We argue that the random coil description – being a mean-field statistical model – is only appropriate when local ordering of the chains does not play a role. Hence care should be taken when applying it to interpret experimental observations on mesophase formation. In contrast the folded-chain fringed-micelle model starts from local correlations. If these are short-range, then the models are not in conflict with each other but complementary. If the local correlations are more extended, the disordered melt is at some stage not homogeneous anymore. Hence the crucial question is for which scale each of the models is relevant and how the transition should be made.

We reconsidered the folded-chain fringed-micelle model, proposed nearly forty years ago, and found it to be appropriate to explain mesophase ordering and crystallisation in the polymer melt and amorphous state. Putting together the evidence provided by Strobl for crystallisation as a multi-stage process [13], the folded-chain fringed-micellar grain model [202–206], the smectic phase of iPP [6, 152, 153], density fluctuation before crystallization [17–29], lozengic and butterfly SANS patterns [217–221] and recent computer simulation [229, 230], the similarity of the resulting pictures may not be accidentally. In order to achieve a deeper understanding of polymer crystallization, the polymer community perhaps has to reconsider the old question: What is the polymer melt? Flory and Uhlmann suggested in 1979 that 'the time has come that controversies on the morphology of amorphous polymers should be laid to rest', but this moment might still not have come yet [234].

Acknowledgements The authors thank Prof. Gregory S.Y. Yeh (Michigan, USA) and Dr. Jan Groenewold (Delft, The Netherlands) for valuable discussions. This work is part of the Softlink research program of the 'Stichting voor Fundamenteel Onderzoek der Materie' (FOM), which is financially supported by the 'Nederlandse Organisatie voor Wetenschappelijk Onderzoek' (NWO).

References

1. Nakatani AI, Dadmun MD (1995) Flow-induced Structure in Polymers. American Chemical Society
2. Miller RL (1979) Flow-induced Crystallization in Polymer Systems. Gordon and Breach, New York
3. Gurra G, Titomanlio G, Meisel I, Grieve K, Kniep CS, Spiegel S (2002) Flow-induced Crystallization of Polymers. John Wiley & Sons
4. Kumaraswamy G, Issian AM, Kornfield JA (1999) Macromolecules 32:7537
5. Kornfield JA, Kumaraswamy G, Issaian AM (2002) Ind Eng Chem Res 41:6383
6. Li LB, de Jeu WH (2004) Faraday Discuss 128:in press
7. Muthukumar M (2003) Phil Trans R Soc Lond A 361:539
8. Armitstead K, Goldbeck-Wood G (1992) Adv Polym Sci 100:221
9. Sadler DM, Gilmer GH (1986) Phys Rev Lett 56:2708
10. Point JJ (1979) Macromolecules 12:770
11. Lauritzen JL, Hoffman JD (1960) J Res Natl Bur Stand 64:73
12. Hoffman JD, Miller RL (1997) Polymer 38:3151
13. Strobl G (2000) Eur Phys J E 3:165
14. Cheng SZD, Li CJ, Zhu L (2000) Eur Phys J E 3:195
15. Lotz B (2000) Eur Phys J E 3:185
16. Muthukumar M (2000) Eur Phys J E 3:199
17. Imai M, Mori K, Mizukami T, Kaji K et al. (1992) Polymer 33:4451
18. Imai M, Kaji K, Kanaya T (1993) Phys Rev Lett 71:4162
19. Imai M, Kaji K, Kanaya T (1994) Macromolecules 27:7103
20. Iami M, Kaji K, Kanaya T (1995) Phys Rev B 52:12696
21. Matsuba G, Kanaya T, Saito M et al (2000) Phys Rev E 62:R1497
22. Ryan AJ et al (1999) Faraday Discuss 112:13
23. Terrill NJ, Fairclough PA, Andrews ET, Komanschek BU, Young RJ, Ryan AJ (1997) Polymer 39:2381
24. Olmsted PD, Poon WCK, Mcleish TCB et al.(1998) Phys Rev Lett 81:373
25. Heeley EL et al (2002) Faraday Discuss 122:343
26. Heeley EL et al (2003) Macromolecules 36:3656

27. Ezquerra TA, Lopez-Cabarcos E, Hsiao BS, Balta-Calleja FJ (1996) Phys Rev E 54:989
28. Matsuba G, Kaji K, Nishida K et al(1999) Macromolecules 32:8932
29. Matsuba G, Kaji K, Nishida K (1999) Polym J 31:722
30. Katayama K, Amano T, Nakamura K (1968) Kolloid Z Z Polym 226:125
31. Yeh GSY, Geil PH (1967) J Macromol Sci Phys B 1:235
32. Yeh GSY, Geil PH (1967) J Macromol Sci Phys B 1:251
33. Schultz JM, Lin JS, Hendricks RW, Peterman J, Gohil JM (1981) J Polym Sci Polym Phys Ed 19:609
34. de Gennes PG, Pincus P (1977) Polym Preprints 18:161
35. Kim YH, Pincus P (1979) Biopolymers 18:2315
36. Matheson R, Flory PJ (1984) J Phys Chem 88:6606
37. Vertogen G, de Jeu WH (1988) Thermotropic Liquid crystals: Fundamentals. Springer, Berlin
38. de Gennes PG (1996) The Physics of Liquid crystals. Clarendon
39. Wunderlich B, Grebowicz J (1984) Adv Polym Sci 60:1
40. Wunderlich B, Moller M, Grebowicz J, Baur H (1988) Adv Polym Sci 87:1
41. Allegra G, Meille SV (2004) Macromolecules 37:3487
42. Ciffrri A, Krigraum WR, Meyer RB (1982) I. Academic Press, New York
43. Donald AM, Windle AH (1992) Liquid Crystalline Polymers. Cambridge University Press, New York
44. Hamley IW (2000) The Physics of Block Copolymers. Oxford University Press, New York
45. Smith GW (1975) in: Brown GH (ed) Advances in Liquid Crystals vol.1. Academic Press, New York, p. 193
46. Ungar G (1993) Polymer 34:2050
47. Wunderlich B (1999) Thermochimica Acta 340/341:37, (1997) Macromol Symp 113:51
48. Flory PJ (1956) Proc Roy Soc London A 234:73
49. Onsager L (1949) Ann N Y Acad Sci 5:627
50. Maier W, Saupe A (1958) Z Naturforsch A 13:564
51. Plotkin SS, Onuchic JN Quarterty (2002) Reviews of Biophysics 35:111
52. Poland D, Scheraga HA (1970) Theory of Helix-Coil Transitions in Biopolymers. Academic, New York
53. Olmsted PD, Goldbart PM (1992) Phys Rev A 46:4966
54. Bruinsma RF, Safinya CR (1991) Phys Rev A 43:5377
55. Fischer E, Callaghan PT (2001) Phys Rev E 64:011501
56. Lenstra TAJ, Dogic Z, Dhont JKG (2001) J Chem Phys 114:10151
57. Kuhn W, Grun F (1942) Kolloid-Z 101:248
58. Chandrasekhar S (1943) Rev Mod Phys 15:1
59. Kratky O, Porod G (1949) Rec Trav Chim 68:1106
60. Doi M, Edwards SF (1986) The Theory of Polymer Dynamics. Oxford University Press, Oxford
61. Schroeder CM, Babcock HP, Shaqfeh ESG, Chu S (2003) Science 301:1515
62. Li HB, Zhang WK, Xu WQ, Zhang X (2000) Macromolecules 33:465
63. Buhot A, Halperin A (2000) Phys Rev Lett 84:2160
64. Nagai K (1961) J Chem Phys 34:887
65. Tsmashiro MN, Pincus P (2001) Phys Rev E 63:021909
66. Carri GA, Muthukumar M (1999) Phys Rev Lett 82:5405
67. Brochard-Wyart F, de Gennes PG (1988) C R Acad Sci Paris 306:699
68. Flory PJ (1986) Principles of Polymer Chemistry. Cornell Univ Press

69. de Gennes PG (1980) Scaling Concepts in Polymer Physics. Cornell University Press, Ithaca, NY
70. de Gennes PG (1971) J Chem Phys 55:572
71. Mandelkern L (1956) Crystallization of Polymers. McGraw-Hill, New York
72. Flory PJ (1956) J Am Chem Soc 78:5222
73. Godovsky YK, Valetskaya LA (1991) Polym Bull 27:221
74. Ostwalds W (1897) Z Phys Chem 22:286
75. Sutton M et al (1989) Phys Rev Lett 62:288
76. Ten Wolde PR, Ruiz-Montero MJ, Frenkel D (1995) Phys Rev Lett 75:2714
77. Shen YC, Oxtoby DW (1996) Phys Rev Lett 77:3585
78. Hikosaka M et al (1992) J Macromol Sci B 31:87
79. Keller A, Cheng SZD (1998) Polymer 39:4461
80. Sirota EB, Herhold AB (1999) Science 283:529
81. ten Wolde PR, Frenkel D (1997) Science 277:1975
82. Hedden RC, Tachibana H, Duncan TM, Cohen C (2001) Macromolecules 34:5540
83. Callaghan PT, Kilfoil Ml, Samulski ET (1998) Phys Rev Lett 81:4524
84. Chu B, Hsiao BS (2001) Chem Rev 101:1727
85. Wunderlich B (2001) Prog Polym Sci 28:383
86. Tadokoro H (1979) Structure of Crystalline Polymers. John Wiley & Sons Inc
87. Jog JP (1995) J Macromol Sci Rev Macromol Chem Phys C 35:531
88. Li LB, Huang R, Lu A (2000) Polymer 41:6943
89. Shimada T, Doi M, Okano K (1988) J Chem Phys 88:2815
90. Doi M, Shimada T, Okano K (1988) J Chem Phys 88:4070
91. Gonzalez C, Zamora F, Guzaman GM, Quimica-Fisica LM (1987) J Macromol Sci B 26:257
92. Bonart von R (1966) Kolloid Z Z Polym 213:1
93. Bonart von R (1968) Kolloid Z Z Polym 231:16
94. Asano T, Seto T (1973) Polym J 5:72
95. Asano T, Balta-Calleja FJ, Flores A, Tanigaki M, Mina MF, Sawatari C, Itigaki H, Takahahi H, Hatta I (1999) Polymer 40:6475
96. Ran S, Wang Z, Burger C, Chu B, Hsiao BS (2002) Macromolecules 35:10102
97. Keum JK, Kim J, Lee SM, Song HH, Son YK, Choi JI, Im SS (2003) Macromolecules 36:9873
98. Blundell DJ, Mahendrasingam A, Martin C, Fuller W, MacKerron DH, Harvie JL, Oldman RJ, Riekel C (2000) Polymer 41:7793
99. Mahendrasingam A, Blundell DJ, Martin C, Fuller W, MacKerron DH, Harvie JL, Oldman RJ, Riekel C (2000) Polymer 41:7803
100. Mahendrasingam A, Martin C, Fuller W, Blundell DJ, Oldman RJ, MacKerron DH, Harvie JL, Riekel C (2000) Polymer 41:1217
101. Blundell DJ, Mahendrasingam A, Martin C, Fuller W (2000) J Mater Sci 35:5057
102. Blundell DJ, Mackerron DH, Fuller W, Mahendrasingam A, Martin C, Oldman RJ, Rule RJ, Riekel C (1996) Polymer 37:3303
103. Mahendrasingam A, Martin C, Fuller W, Blundell DJ, Oldman RJ, Harvie JL, Mackerron DH, Riekel C, Engstrom P (1999) Polymer 40:5553
104. Kawakamia D, Hsiao BS, Ran S, Burger C, Fu B, Sics I, Chu B, Kikutani T (2004) Polymer 45:905
105. Kawakami D, Ran S, Burger C, Fu B, Sics I, Chu B, Hsiao BS (2003) Macromolecules 36:9275
106. Fukao K, Koyama A, Tahara D, Kozono Y, Miyamoto Y, Tsurutani N (2003) J Macromol Sci Phys B 42:717

107. Auriemma F, Corradini P, De Rosa C, Guerra G, Petroccone V (1992) Macromolecules 25:2490
108. García Gutiérrez MC, Karger-Kocsis J, Riekel C (2002) Macromolecules 35:7320
109. Carr PL, Nicholson TM, Ward IM (1997) Polym Adv Tech 8:592
110. Welsh GE, Blundell DJ, Windle AH (1998) Macromolecules 31:7562
111. Welsh GE, Blundell DJ, Windle AH (2000) J Mater Sci 35:5225
112. Lotz B, Wittmann JC, Lovinger AJ (1996) Polymer 37:4979
113. Natta G, Peraldo M, Corradini P (1956) Rend Accad Naz Lincei 26:14
114. Gailey JA, Ralston PH (1964) Plast Eng Trans 4:29
115. Miller RL (1960) Polymer 1:135
116. Bruckner S, Meille SV, Petraccone V, Pirozzi B (1991) Prog Polym Sci 16:361
117. Hosemann R (1951) Acta Crystallogr 4:520
118. Zannetti R, Celotti G, Armigliato A (1970) Eur Polym J 6:879
119. Wycoff HW (1962) J Polym Sci 62:83
120. Gomez MA, Tanaka H, Tonelli E (1987) Polymer 28:2227
121. Corradini P, Petraccone V, De Rosa C, Guerra G (1986) Macromolecules 19:2699
122. Corradini P, De Rosa C, Guerra G, Petraccone V, (1989) Polym Comm 30:281
123. Poncinn-Epaillard F, Brosse JC, Falher T (1997) Macromolecules 30:4415
124. Martorana A, Piccarolo S, Sapoundjieva D (1999) Macromol Chem Phys 200:531
125. Miyamoto Y, Fukao K, Yoshida T, Tsurutani N, Miyaji H (2000) J Phys Soc Jap 69:1735
126. Wang ZG, Hsiao BS, Srinivas S, Brown GM, Tsuo AH, Cheng SZD, Stein RS (2001) Polymer 42:7561
127. Li ZM, Li LB, Yang W, Yang MB, Huang R (2004) Macromol Rap Comm 25:533
128. Monasse B (1992) J Mater Sci 27:6047
129. Liedauer S, Eder G, Janeschitz-Kriegl H (1993) Intern Polym Proc 8:236
130. Jerschow P, Janeschitz-Kriegl H (1997) Intern Polym Proc 12:72
131. Vleeshouwers S, Meijer HEH (1996) Rheol Acta 35:391
132. Hosier IL, Bassett DC, Moneva IT (1995) Polymer 36:4197
133. Jay F, Haudin JM, Monasse B (1999) J Mater Sci 34:2089
134. Chen LM, Shen KZ (2000) J Appl Polym Sci 78:1906
135. Duplay C, Monasse B, Haudin J et al (1999) Polym Inter 48:320
136. Agarwal PK et al (2003) Macromolecules 36:5226
137. Somani RH et al (2002) Macromolecules 35:9096
138. Somani RH, Yang L, Hsiao BS (2002) Physica A 304:145
139. Ran SF, Zong XH, Fang DF, Hsiao BS, Chu B, Philips RA (2001) Macromolecules 34:2569
140. Somani RH, Hsiao BS, Nogales A et al (2000) Macromolecules 33:9385
141. Somani RH et al (2001) Macromolecules 34:5902
142. Kumaraswamy G, Issian AM, Kornfield JA (1999) Macromolecules 32:7537
143. Kmaraswamy G, Verma RK, Issian AM et al (2000) Polymer 41:8931
144. Kumaraswamy G, Kornfield JA, Yeh F, Hsiao BS (2002) Macromolecules 35:1762
145. Kornfield JA, Kumaraswamy G, Issaian AM (2002) Ind Eng Chem Res 41:6383
146. Seki M, Thurman DW, Oberhauser JP, Kornfield JA (2002) Macromolecules 35:2583
147. Elmoumni A, Winter HH, Eaddon AJ, Fruitwala H (2003) Macromolecules 36:6453
148. Nogales A, Mitchell GR, Vaughan AS (2003) Macromolecules 36:4898
149. Huo H, Jiang S, An L, Feng J (2004) Macromolecules 37:2478
150. Zhu PW, Edward G (2004) Macromolecules 37:2658
151. Li LB, de Jeu WH (2003) Macromolecules 36:4862
152. Li LB, de Jeu WH (2004) Phys Rev Lett 92:075506

153. Zhu XY, Yan DY, Fang YP (2001) J Phys Chem B 105:12461
154. Kaganer VM, Diele S, Ostrovskii BI, Haase W (1997) Mol Mater 9:59
155. Keymeulen HR, de Jeu WH, Slatery JT, Veum M (2002) Eur Phys J E 9:443
156. Minami S, Tsurutani N, Miyaji H, Fukao K, Miyamoto Y (2004) Polymer 45:1429
157. Lee C, Johannson O, Flanigam O, Hahn P (1967) Polym Prepr 10:1311
158. Beatty C, Pochan J, Froix M, Hinman D (1975) Macromolecules 8:547
159. Saxena H, Hedden RC, Cohen C (2002) J Reho 46:1177
160. Godovsky YK, Papkov VS, Magonov SN (2001) Macromolecules 34:976
161. Magonov SN, Elings V, Papkov VS (1997) Polymer 38:297
162. Matveev MM, Sidorkin AS, Klinskikh AF (1997) Phys Solid State 39:329
163. Molenberg A, Moeller M (1997) Macromolecules 30:8332
164. Molenberg A, Moeller M, Sautter E (1997) Prog Polym Sci 22:1133
165. Inomata K, Yamamoto K, Nose T (2000) Polym J 32:1044
166. Molenberg A, Moeller M, Soden Wv (1998) Acta Polym 49:45
167. Tsvankin DYa, Papkov VS, Godovsky YuK, Zhdanov AA (1985) J Polym Sci, Polym Chem Ed 23:1043
168. Out G, Turestskii AA, Snijder M, Moeller M, Papkov VS (1995) Polymer 36:3213
169. Godovsky YuK (1992) Angew Makromol Chem 202/203:187
170. Papkov VS, Godovsky YuK, Svistunov VS, Zhdanov AA (1989) Vysokomol Soedin A 31:1577
171. Godovsky YuK, Papkov VS (1988) Adv Polym Sci 88:129
172. Ganicz T, Stanczyk WA (2003) Prog Polym Sci 28:303
173. Batra A, Hedden RC, Schofield P, Barnes A, Cohen C, Duncan TM (2003) Macromoles 36:9458
174. Koerner H, Luo Y, Li X, Cohen C, Hedden RC, Ober CK (2003) Macromolecules 36:1975
175. Hedden RC, Saxena H, Cohen C (2000) Macromolecules 33:8676
176. Hedden RC, McCaskey E, Cohen C, Duncan TM (2001) Macromolecules 34:3285
177. De Gennes PG (1975) C R Acad Sci Ser B 281:101
178. Warner M, Wang XJ (1991) Macromolecules 24:4932
179. Mark JE, Chin DS, Su TK (1978) Polymer 19:407
180. Miller KJ, Grebowicz J, Wesson JP, Wunderlich B (1990) Macromolecules, 23:849
181. Pennings AJ, Zwijnenburg AJ (1979) J Polym Sci Polym Phys 17:1011
182. Strobl G (1997) The Physics of Polymers: Concepts for Understanding Their Structures and Behavior. Springer Verlag, Berlin
183. Wong GCL, de Jeu WH, Shao H, Liang KS, Zentel R (1997) Nature 389:576
184. Pennings AJ (1967) Crystal Growth
185. Hobbs JK, Homphris ADL, Miles MJ (2001) Macromolecules 34:5508
186. Keller A, Kolnaar H (1997) Materials Science and Technology. A Comprehensive Treatment. Ch 4, 18, p 189–268
187. Liu TX, Tjiu WC, Petermann J (2002) J Cryst Grow 243:218
188. Rueda DR, Ania F, Balta Calleja FJ (1997) Polymer 38:2027
189. Monks AW, White HM, Bassett DC (1996) Polymer 37:5933
190. Shimizu J, Kikutani T, Takaku A, Okui N (1984) Seni Gakkaishi 40:T177
191. Wu J, Schultz JM, Yeh F, Hsiao BS, Chu B (2000) Macromolecules 33:1765
192. Hieber CA (1995) Polymer 36:1455
193. Papkov VS, Godovsky YK, Litvinov VM, Zhdanov AA (1984) J Polym Sci, Polym Chem Ed 22:3617
194. Luch D, Yeh GSY (1973) J Appl Phys 43:4326
195. Toki S, Hsiao BS (2003) Macromolecules 36:5915

196. Trabelsi S, Albouy PA, Rault J (2003) Macromolecules 36:7624
197. Miyamoto Y, Yamao H, Sekimoto K (2003) Macromolecules 36:6462
198. Yeh GSY, Hong HZ (1979) Polym Eng Sci 19:395
199. Coppola S, Grizzuti N, Maffettone PL (2001) Macromolecules 34:5030
200. Dukovski I, Muthukumar M (2003) J Chem Phys 118:6648
201. Doufas AK, McHugh AJ, Miller C (2000) J Non Newton Fluid Mech 72:27
202. Wunderlich B (1970) Macromolecular Physics, vol 1. Academic Press, New York
203. Flory PJ (1969) Statistical Mechanics of Chain Molecules. Interscience, New York
204. Yeh GSY (1972) J Macromol Sci Phys B 6:451
205. Yeh GSY (1972) Pure Appl Chem 31:65
206. Geil PH (1976) J Macromol Sci Phys B 12:173
207. Pechhold W, Blasenbrey S (1970) Kolloid Z u Z Polymere 241:955
208. Flory PJ (1979) Faraday Discuss 68:14
209. Fischer EW (1978) J Non Cry Solids 31:181
210. Flory PJ (1975) Macromol Chem 8:1
211. Xu Z, Hadjichristidis N, Fetters LJ, Mays JW (1995) Adv Polym Sci 120:1
212. Smith GD, Yoon DY, Jaffe RL, Colby RH, Krishnamoorti R, Fetters LJ (1996) Macromolecules 29:3462
213. Ballard DGH, Cheshire P, Longman GW, Schelten J (1978) Polymer 19:379
214. Ballard DGH, Burgess AN, Nevin A, Cheshire P, Longman GW, Schelten J (1980) Macromolecules 13:677
215. Gupta MR, Yeh GSY (1978) J Macromol Sci Phys B 15:119
216. Wang CS, Yeh GSY (1978) J Macromol Sci Phys B 15:107
217. Richter D, Springer T (ed) (1987) Polymer Motion in dense Systems. Springer-Verlag, Berlin, page 86,104,112
218. Mendes E, Lindner JP, Buzier M, Boué F, Bastide J (1991) Phys Rev Lett 66:1595
219. Geissler E, Horkay F, Hecht A (1993) Phys Rev Lett 71:645
220. Rouf-George C, Munch JP, Schosseler F, Pouchelon A, Beinert G, Boue F, Bastide J (1997) Macromolecules 30:8344
221. Mendes E, Oeser R, Hayes C, Boue F, Bastide J (1996) Macromolecules, 32:5574
222. Torre R, Bartolini P, Righini R (2004) Nature 428:296
223. Mcalea KP, Schultz JM, Wardner KH, Wignall GD (1987) J Polym Sci Polym Phys Part B 25:651
224. Zachmann HG (1969) Kolloid Z Z Polym 231:504
225. Hermann K, Gerngross O, Abitz W (1930) Z Phys Chem (Munich) B10:371
226. Flory PJ (1962) J Am Chem Soc 84:2857
227. Allegra G (1980) Ferroelectrics 30:195
228. Allegra G, Meille SV (1999) Phys Chem Chem Phys 1:5179
229. Muthukumar M, Welch P (2000) Polymer 41:8833
230. Welch P, Muthukumar M (2001) Phys Rev Lett 87:218302
231. Flory PJ, Yoon DY (1978) Nature 272:226
232. Klein J, Ball R (1979) Faraday Discuss 68:198
233. Hu W, Frenkel D, Mathot VBF (2003) Macromolecules 36:549
234. Uhlmann DR (1979) Faraday Discuss 68:87

Adv Polym Sci (2005) 181: 121–152
DOI 10.1007/b107177

Published online: 30 June 2005

Stepwise Phase Transitions of Chain Molecules: Crystallization/Melting via a Nematic Liquid-Crystalline Phase

Akihiro Abe[1] (✉) · Hidemine Furuya[2] · Zhiping Zhou[3] · Toshihiro Hiejima[1] · Yoshinori Kobayashi[1]

[1]Department of Applied Chemistry, Tokyo Polytechnic University, 1583 Iiyama, Atsugi 243-0297, Japan
aabe@chem.t-kougei.ac.jp

[2]Department of Organic and Polymeric Materials, Tokyo Institute of Technology, 2-12-1-H-128 Ookayama, Meguro, Tokyo 152-8552, Japan

[3]Graduate School, Jiangsu University, 301 Xuefu Road, Zhenjiang, Jiangsu Province 212013, P.R. China

Abstract Segmented chain molecules involving mesogenic units along the backbone chain, conventionally called mainchain liquid crystals, often exhibit an enatiotropic nematic liquid-crystalline phase over a certain temperature range between the crystal and the isotropic melt. The phase transitions involved are usually of first order. Spectroscopic analyses such as X-ray and neutron diffraction, IR and Raman absorption, and ^{1}H and ^{2}H NMR demonstrated that the orientational ordering of mesogenic cores in the mesophase are strongly coupled with the spatial arrangement of the spacer. While the orientation of the molecular axis (i.e., the whole molecule) varies as a function of temperature, the well-defined nematic conformation of the spacer remains almost unaltered over the entire range of the liquid-crystalline state. In these compounds, the conformational transition takes place in the order isotropic random coil ↔ nematic conformation ↔ extended crystalline form, or vice versa during the crystallization/melting process. In this article, we first present a brief summary of the entropy of fusion of chain molecules. In the second part, conformational entropy changes associated with the melting via a nematic phase, or the crystallization via a nematic mesophase, are reviewed. The experimental data on liquid-crystal transitions were mostly collected from studies on oligomeric compounds such as dimer and trimer mainchain liquid crystals with neat chemical structures. Finally it is suggested that further studies on the phase transition of chain molecules via a nematic mesophase may provide some valuable information regarding the embryonic stage of the polymer crystallization mechanism.

Keywords Mainchain liquid crystals · Nematic conformation · Crystallization · Liquid-crystallization · Transition entropies

1 Introduction

The entropy of melting of simple low molar mass compounds is considered to be due largely to the destruction of long-range order. The values are generally rather small and remain in the range 10–40 $\mathrm{J\,mol^{-1}\,K^{-1}}$, e.g., cyclohexane (9.4), carbon tetrachloride (10.2), benzene (34.7), 1,4-dioxane (45.2), isobutane (39.8), with the units being joules per mole per kelvin [1]. The inert gases have entropies of melting which are roughly of the order of the gas constant R, and are sometimes identified as communal entropy [2]. In polymers, there is an additional contribution from the intramolecular rearrangement due to rotational isomerism [3]. Imperfections of the crystalline morphology in polymers often cause melting to occur over a finite temperature range. Various pieces of crystallographic evidence, however, indicate that the melting of crystalline polymers is analogous in its major aspects to the first-order transition of ordinary crystalline compounds [4]. Thermodynamic data for a series of n-alkanes $CH_3(CH_2)_{n-2}CH_3$ have been compiled in the NIST data book [1]. Chain-length dependence of the fusion entropy under ordinary pressure can be elucidated therefrom. By ignoring a small fluctuation arising from the odd–even oscillation with n, the entropy of melting expressed in terms of $\mathrm{J\,mol^{-1}\,K^{-1}}$ tends to increase monotonically with the number of constituent carbon atoms (n) of the chain. For lower-member n-alkanes with $n = 1$–20, the increment per methylene unit is about 10 $\mathrm{J\,mol^{-1}\,K^{-1}}$. The value goes down to approximately 6 $\mathrm{J\,mol^{-1}\,K^{-1}}$ for the range $n = 21$–30, and up again to approximately 8 $\mathrm{J\,mol^{-1}\,K^{-1}}$ for $n = 30$–35. Many members exhibit a premelting transition prior to the true fusion. For the purpose at hand, the entropy changes associated with such pretransitions are ignored in this estimate. According to a careful analysis of Wunderlich and Czornyj [5], the entropy of melting of polyethylene (PE) is estimated to be 9.91 $\mathrm{J\,(mol\,of\,CH_2)^{-1}\,K^{-1}}$ at the equilibrium melting temperature 414.6 K.

It is known for various semiflexible polymers that the configurational entropy change is of primary importance in establishing the melting temperature. The rotational isomeric character of chemical bonds facilitates the statistical mechanical description of chain molecules. The configurational partition function of a given chain can be easily derived within the rotational isomeric state (RIS) approximation [6, 7]. The entropy change at the phase transition may be obtained as the difference between the two relevant states. For a variety of conventional polymers, the entropy change calculated in this manner has been found to be in good agreement with the experimental estimate of the entropy of melting at constant volume. The constant-volume transition entropy required here may be derived from the entropy of fusion measured under constant pressure with proper correction for the volume change. Although the physical basis of such an entropy separation is somewhat controversial [5], the technique is well accepted for long-chain molecules [4, 8–17].

In addition to the ambiguities inherent to the physical concept, the determination of thermodynamic quantities such as the latent heat and the volume change at the transition is often hampered by the fact that the crystalline state of chain molecules is quite complex. The polymer crystals are usually polycrystalline and coexist with the disordered amorphous domain. An accurate estimation of the equilibrium melting temperature defined for a perfectly aligned crystal requires great effort [5, 18, 19]. At the melting temperature, equilibrium usually exists between the liquid and somewhat imperfect crystalline phases.

The uncertainties associated with the transition between the liquid (I) and the crystalline solid phase (C) should be greatly relieved for the equilibrium between the isotropic liquid (I) and the liquid-crystalline (LC) phase. Segmented chain molecules involving mesogenic units along the backbone chain, conventionally called mainchain liquid crystals, often exhibit a nematic LC state (N) over a certain temperature range between the crystal and the isotropic melt [20, 21]. The phase transitions involved are usually of first order. According to pressure–volume–temperature (*PVT*) studies on oligomeric compounds, the volume changes associated with the NI transition are about 10–25% of those observed at the CN boundary. The nematic mesophase should thus be highly fluid, accommodating various conformations of the chain segment unless they are incompatible with the nematic environment. Spectroscopic analyses such as X-ray [22] and neutron diffraction [23], IR and Raman absorption [24], and ^{1}H and ^{2}H NMR [25] demonstrated that the orientational ordering of mesogenic cores is strongly coupled with the spatial arrangement of the spacer [26]. In the nematic LC state, the spacers are comparatively extended, suggesting that they may possibly participate in the anisotropic intermolecular interactions in the field. While the orientation of the molecular axis (i.e., the whole molecule) varies as a function of temperature, the ensemble of the well-defined nematic conformation of the spacer remains almost unaltered over the entire range of the LC state. These results were derived from analyses of oligomeric samples such as dimer and trimer mainchain liquid crystals with neat chemical structures [27]. For a series of segmented chain molecules, these oligomeric compounds have been proven to be a good model to study the thermodynamic role of the flexible spacer in determining the phase transition behavior [28–30].

In this article, we first present a brief summary of the entropy of fusion of chain molecules. In the second part, conformational entropy changes associated with the melting via a nematic phase, or the crystallization via a nematic mesophase, will be reviewed.

2 Role of Conformational Entropy in Determining the Phase Transition Behavior of Chain Molecules

2.1 RIS Calculation of Conformational Entropy of Chain Molecules

Conformational entropy S^{conf} may be defined as

$$S^{\mathrm{conf}} = k \ln Z + kT(\mathrm{d} \ln Z/\mathrm{d}T)\,, \tag{1}$$

where Z represents the configuration partition function,

$$Z = J^* U_1^{(n)} J\,, \tag{2}$$

where J^* and J are the row and column vectors, respectively, taking care of the initial and terminal bonds of the chain comprising n bonds [11, 31]. Enumeration of all possible configurations can be performed by the matrix multiplication of the conformational statistical weight matrix U_i defined within the RIS approximation. With U_i properly defined for bond i, calculation of the average of any conformation-dependent properties is straightforward. The usefulness of the conformational partition function thus defined has been demonstrated by a vast majority of successful examples: various properties of polymer chains have been studied in dilute solution as well as in the bulk molten state by both experimental and theoretical techniques [6, 7, 32].

The conformational entropy defined by Eq. 1 is directly related to the flexibility or rigidity of given polymer chains. Thermodynamic properties of the bulk state are known to be largely affected by the flexibility of the polymer chain. The RIS information is fundamentally important in predicting bulk properties of polymers from a given first-order structure. A great deal of effort has been made to elucidate the role of conformational entropy in determining the phase transition behavior of chain molecules.

2.2 Identification of Conformational Transition Entropy by the *PVT* Method

As illustrated in Fig. 1, the mechanism involved in the melting of crystals is markedly different between polymeric and monomeric globular compounds. While the destruction of long-range order, i.e., the randomization of centers of gravity, is the major contribution to the entropy of melting in low molecular weight compounds, changes in the spatial arrangement of polymeric chains are considered to be the major factor in other compounds. The role of rotational isomerism on the crystallization/melting behavior has thus been an important issue in the history of polymer science [4, 18, 19]. In the molten amorphous state, chain molecules are assumed to take random-coil arrangements unperturbed by the long-range interactions along the chain, i.e., $Z = Z_{\mathrm{I}}$ [33], while the chains are confined in a well-ordered single conformation in the crystalline state; thus, $Z = 1$. The conformational entropy change S^{conf} can be easily calculated by taking the difference between these two states (Eq. 1). Strictly speaking, however, the validity of the treatment is guaranteed only for an isolated single chain. The crystallization/melting under constant pressure usually accompanies some significant volume change of the system. In order to estimate the conformational contribution to the latent entropy, the value observed under the isobaric condition must be corrected for the volume change.

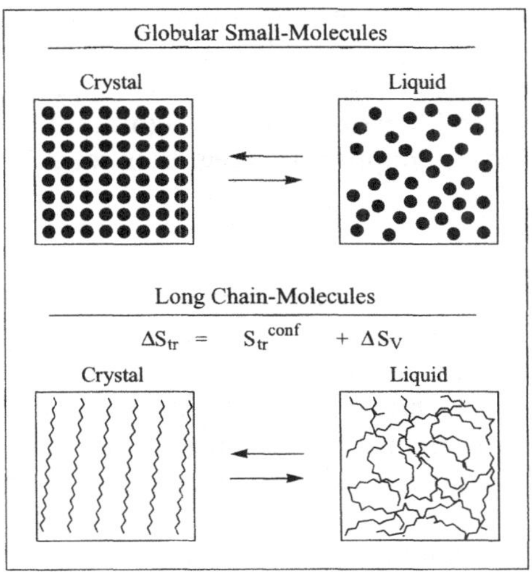

Fig. 1 Illustration of the crystallization/melting process of globular small molecules and long polymeric molecules. The latent entropy of the latter transition (ΔS_{tr}) consists of two major contributions: the configurational term ($S_{\mathrm{tr}}^{\mathrm{conf}}$) and the volume-dependent term (ΔS_V) (see text)

According to Mandelkern's formulation [4], the increase in entropy due to the volume expansion at the melting point can be conventionally estimated by

$$\Delta S_V = (\alpha/\beta)\Delta V = \gamma \Delta V \,, \tag{3}$$

where α, β, and γ, respectively, denote the thermal expansion coefficient, the isothermal compressibility, and the thermal pressure coefficient.

$$\gamma = (\partial P/\partial T)_V = (\partial S/\partial V)_T \,. \tag{4}$$

The constant-volume entropy change $(\Delta S_{\text{tr}})_V$ at the melting point can then be derived from the relation

$$(\Delta S_{\text{tr}})_V = (\Delta S_{\text{tr}})_P - \Delta S_V \,, \tag{5}$$

where tr denotes the melting transition and $(\Delta S_{\text{tr}})_P$ is the melting entropy observed under atmospheric pressure. The values of $(\Delta S_{\text{tr}})_V$ have been estimated for a variety of polymers and have often been shown to be in reasonable accord with those of the conformational entropy calculated according to the RIS approximation [4, 8–17], suggesting that a major contribution to the melting entropy arises from the internal rotation around the constituent bonds.

In the real system, volume expansion or contraction takes place at the transition owing to the density difference between the two phases in equilibrium. The entropy separation according to Eqs. 3–5 is a hypothetical process assuming that the volume of the isotropic fluid may be compressed to that of the solid state without affecting the configurational part of the entropy of the chain molecules. The validity of such an assumption has been questioned by several authors [5, 17, 34–37]. Wunderlich et al. pointed out that the volume dependence of γ during the compression from the liquid to the solid volume may not be negligible, and thereby leads to a significant underestimate of the ΔS_V term. They proposed adopting an integration form such as

$$\Delta S_V = \int \gamma(V)\, \mathrm{d}V \tag{6}$$

to replace Eq. 3. To what extent the entropy of the system including the configuration of chain molecules could be affected by the compression of the bulk volume is apparently an important point which must be carefully investigated. The phase rule, however, dictates that the direct determination of the $\gamma(V)$ function at the melting point is impossible. At the phase boundary, a slight increase in pressure could immediately cause crystallization at this point.

Since the values determined for γ by extrapolation from the liquid state (γ_l) and the crystalline state (γ_c) usually do not coincide, the two pathways to estimate ΔS_V by Eq. 3 inevitably yield unequal values [15]. The discrepancy arising from these two different pathways will be discussed later for a specific example. In practice, the extrapolation from the fluid state is more easily carried out. The variation of the spatial configuration of chain molecules by compression of the volume has been examined by a molecular dynamics (MD) simulation in conjunction with the estimate of the γ–volume relation [38]. This subject will be discussed in Sect. 2.2.2.

2.2.1 Thermal Pressure Coefficient $\gamma = (\partial P/\partial T)_V$ of Isotropic Liquids

As stated already, the value of γ to be used in Eq. 3 or Eq. 6 is critically important in the estimation of ΔS_V. The variation of γ with specific volume V_{sp} has been calculated according to the relation $\gamma = \alpha/\beta$ at given temperatures from the *PVT* data for methane [39, 40], ethane [41, 42], *n*-undecane [43, 44], and PE (DP = 9000) [43] for the liquid state [45]. In all these systems, non-polar van der Waals type intermolecular (intersegmental) interactions prevail. Figure 2 illustrates an overview of the γ vs. V_{sp} plots for the series of *n*-alkanes mentioned earlier. In each molecular system, the γ–V_{sp} curve tends to shift slightly as a function of temperature: variations become marked at higher pressure (smaller V_{sp}). The γ values tend to be enhanced as the specific volume decreases at given temperatures. It is interesting to note that the γ–V_{sp} behaviors are more or less alike for all liquids, so they can be shifted to form a master curve, which reminds us that a simple van der Waals fluid obeys the relation $\gamma = R/(V - b)$, with b being the van der Waals parameter. In this respect, the series of chain molecules mentioned earlier are not much different from small globular particles in their γ–V_{sp} behaviors.

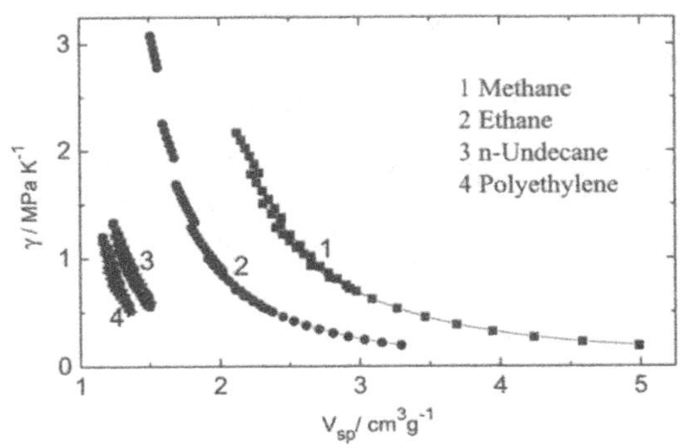

Fig. 2 An overview of the $\gamma - V_{sp}$ curves for a series of *n*-alkanes, estimated by the analysis of *PVT* data reported in the literature. Calculations were carried out for given temperatures: methane [39, 40] 100–190 K with 20 K intervals, ethane [41, 42] 100–300 K with 40 K intervals, *n*-undecane [43, 44] 313–493 K with 20 K intervals, and polyethylene [43] 413–533 K with 20 K intervals. For simplicity, details are not indicated in the diagram

2.2.2
Elucidation of Constant-Volume Transition Entropy

Representative examples are shown in Table 1, where CI is used to specifically designate the biphasic boundary of the crystal/isotropic melt. The values of γ_{CI} at the melting temperature were estimated by two different methods, and they are distinguished by using parentheses. In method 1, the γ_{CI} values are obtained by the extrapolation from the liquid state to the melting temperature under atmospheric pressure. In method 2, the γ_{CI} values in parentheses are those estimated from the $\gamma - V_{sp}$ curve for T_{CI} in the following manner: the value corresponding to the midpoint V_{sp} between the liquid and the crystal is taken to be the γ_{CI} value at T_{CI}. Let us recall here that in our hypothetical process, the volume of an isotropic liquid must be compressed to that of the solid state without affecting the conformational freedom of the individual molecules. While the former γ_{CI} values correspond to those at the initial point of the compression process (at zero pressure), the latter may be un-

Table 1 Thermal pressure coefficients γ, transition volumes ΔV, and volume-dependent transition entropies ΔS_V of *n*-alkanes and polyoxyethylene (*POE*) for the crystal–isotropic (CI) phase transitions: conformational entropy changes S^{conf} estimated by the rotational isomeric state approximation are included for comparison

Compound	T_{CI} (K)	γ_{CI} (MPa K^{-1})	ΔV_{CI} (cm^3 mol^{-1})	ΔS_V (J mol^{-1} K^{-1})	$(\Delta S_{CI})_P$ (J mol^{-1} K^{-1})	$(\Delta S_{CI})_V$ (J mol^{-1} K^{-1})	S^{conf} (J mol^{-1} K^{-1})
Methane	90.67[a]	2.0 (2.2)[b]	2.66[c]	5.32 (5.85)	10.3[b]	4.98 (4.45)	
n-Undecane	247.6[a,d]	1.3 (1.5)[e]	14.98[d]	19.5 (22.5)	89.6[a]	70.1 (67.1)	61.9[f]
PE	414.6[g]	0.83 (1.2)[e]	3.82[g]	3.17 (4.58)	9.91[g]	6.74 (5.33)	7.41[h]
POE	341.1[i]	1.5 (1.8)[e]	5.1[j]	7.65 (9.18)	27.2[i]	19.6 (18.0)	20.9[l]

For polymers, volumes and entropies are expressed in terms of the number of moles of a repeating unit: $-CH_2-$ for polyethylene (*PE*) and $-OCH_2CH_2-$ for POE. γ values were estimated in two ways: those obtained by method 2 and related transition entropies are shown in *parentheses*.
[a]Ref. [1]
[b]Ref. [39, 40, 45]
[c]Ref. [46]
[d]Ref. [43, 44, 47]
[e]Ref. [43, 45]
[f]Ref. [38]
[g]Ref. [5]
[h]Ref. [50]
[i]Ref. [48]
[j]Ref. [49]
[l]Ref. [16]

derstood as those at the midpoint of the process (under a significantly high pressure). The first method has been conventionally adopted in most studies reported in the literature. The latter method was examined to comply with the comments raised in conjunction with Eq. 6. For the purpose of comparison, the calculated results obtained by using the γ_{CI} values from method 2 are all accommodated in parentheses (Table 1).

The γ_{CI} value from the first method ranges from 2.0 MPa K^{-1} at 90.67 K for methane to 0.83 MPa K^{-1} at 414.6 K for PE. The value (γ_{CI}) was estimated to be 1.5 MPa K^{-1} at 341.1 K for polyoxyethylene (POE). As shown in Fig. 2, the γ values more or less tend to be enhanced as V_{sp} decreases. Consequently, the γ_{CI} values estimated by method 2 (in parentheses) are somewhat higher than those corresponding to method 1: the increments range from 10 (methane) to 45% (PE).

The experimental data required in the estimation of the constant-volume transition entropies $(\Delta S_{CI})_V$ as prescribed in Eqs. 3–5 were collected from the literature for methane [1, 39, 40, 46], *n*-undecane [1, 43, 44, 47], PE [5, 43], and POE [43, 48, 49]. The values of ΔS_V and $(\Delta S_{CI})_V$ thus estimated are shown in columns 5 and 7 in Table 1, where the entropies for the polymers are expressed in terms of a repeating unit. PE is a well-studied polymer, which melts at 414.6 K with a volume change of 0.273 cm^3 g^{-1}, and the associated entropy of fusion is reported to be 9.91 J mol^{-1} K^{-1} [5]. The γ value estimated at the midpoint of the phase transition region (V_{sp} = 1.14 cm^3 g^{-1}) (Fig. 2) is 1.2 MPa K^{-1} at the melting temperature [45]. Use of the γ values obtained by methods 1 and 2, respectively, leads to $(\Delta S_{CI})_V = 6.7$ and 5.3 J (mol of CH$_2$)$^{-1}$ K^{-1}, which should be compared with the conformational entropy [$S_{CI}^{conf} \approx 7$ J (mol of CH$_2$)$^{-1}$ K^{-1}] variously estimated within the RIS approximation [5, 50]. In general, the values of $(\Delta S_{CI})_V$ corresponding to method 1 are larger than those of method 2 (Table 1): 12% (methane), 4% (*n*-undecane), 26% (PE), and 9% (POE). The large discrepancy in PE is due to a rapid rise of the γ–V_{sp} curve in the transition region (Fig. 2). In the other three examples, the effect arising from the choice of the γ_{CI} values is found to be rather small in the final results.

The conformational entropies S_{CI}^{conf} obtained by the RIS calculation using Eqs. 1 and 2 are listed in the last column of the table [16, 38, 50]. The quantities listed in the seventh and eighth columns are favorably compared with each other. It should be interesting to pursue the origin of the residual entropy $(\Delta S_{CI})_V = 4$–5 J mol^{-1} K^{-1} of methane, which is in principle free from any conformational contributions [51]. This is a separate task, however.

As mentioned earlier, *n*-undecane exhibits a premelting transition at about 10 degrees below the melting temperature. The entropy change due to this solid-state (rotor phase) transition is reported to be 27–29 J mol^{-1} K^{-1} [1]. The rotational isomeric (conformational) character of individual chains remains unperturbed by these low-temperature transitions. The physical meaning of this entropy should cause controversial arguments for and against the aforementioned treatment [17, 35, 37]. In Table 1, the $\Delta(S_{\rm CI})_p$ values directly observed at the melting point are adopted [1]. To obtain further insight into the problem, MD simulations were performed for *n*-undecane [38]. The molecule comprises ten skeletal C – C bonds, or eight rotatable bonds. The experimental isobars were well reproduced by the simulation, and accordingly the γ vs. $V_{\rm sp}$ relations for given temperatures derived from the simulation were favorably compared with those obtained from the experimental *PVT* data. The entropy derived from Eq. 5 is $(\Delta S_{\rm CI})_V = 71.6\,\mathrm{J\,mol^{-1}\,K^{-1}}$, in favorable agreement with $S_{\rm CI}^{\rm conf} = 61.9\,\mathrm{J\,mol^{-1}\,K^{-1}}$ [38].[1] The configurational characteristics of the *n*-undecane molecule were investigated for representative ensembles registered in the MD simulation. The spatial configuration of *n*-undecane is not significantly affected by the compression of the bulk volume within the range (below 200 MPa) examined. The *trans–gauche* transition around the C – C bond occurs quite frequently as long as temperatures are kept sufficiently high. At lower temperatures, however, the conformational transition tends to deviate from the ergodic behavior. When a liquid is compressed to the crystalline volume, internal rotations are restricted, and perhaps bond lengths and bond angles are also changed. The estimation of the $\gamma_{\rm CI}$ value may therefore be affected by the pathway adopted in the extrapolation from those of the less dense liquids. What we can do at best is to find a hypothetical process in which the volume of the liquid can be compressed to that of the crystal without creating too many intermolecular conflicts: the intramolecular bond rotation and thus the rearrangement of the spatial configuration of the chain should be permitted even when the entire volume is reduced. In this respect, the results of the MD simulation mentioned previously cast some doubt on the eligibility of the $\gamma_{\rm CI}$ values obtained by method 2. Experimental determination of the configuration entropy for the crystal–isotropic transition is inevitably approximate.

The present examples have provided a useful test for the validity of the hypothetical process adopted conventionally for the estimation of the constant-volume transition entropy. Although the estimation of the $\gamma_{\rm tr}$ value using the relation $\gamma = \alpha/\beta$ under ordinary pressure (method 1) is simple, the process should not be excluded because of its simplicity unless a more prevailing form of $\gamma(V)$ is known.

[1] Wurflinger's view [17] that the $\Delta S_{\rm CI}$ term should comprise all contributions due to all transitions between the crystal and the melt offers an alternative approach in the present arguments. Inclusion of the rotor phase transitions accordingly leads to a larger discrepancy.

3
Crystallization/Melting of Chain Molecules via a Nematic Mesophase

For the purpose of this article, we focus our attention on the nematic mesophase: smectic orders are more crystal-like and thus are beyond our scope. Typical nematic liquid crystals are characterized by a uniaxial order, though imperfect, along the preferred axis of the domain. No such long-range order exists in directions transverse to the domain axis. In most examples, low molar mass (monomer) liquid crystals carry flexible tails. Conformational ordering of these tails in the mesophase has been extensively studied in relation to the odd–even character of the phase behavior with the number of constituent atoms of the pendant chain. Various statistical models and theories have been presented [52–57]. In most cases, however, the ordering of the tail is relatively weak [58, 59].

Phase diagrams have been constructed for several binary mixtures comprising flexible chains and rigid-rod molecules [60–66]. Inclusion of flexible components tends to destabilize the uniaxial order of the LC mesophase owing to the entropy demixing principle, leading to a phase diagram consisting of a narrow mesophase, a biphasic gap, and a wide range of an isotropic mixture as a function of concentration. The biphasic lines separating the ordered and disordered isotropic phases vary sensitively with temperature: the slopes of the biphasic boundary curves serve as a measure of the compatibility of nonmesomorphic solutes with the nematic phase [60, 64, 65]. Conformational ordering of chain molecules accommodated in nematic LCs has been examined by various methods. Although interpretations are somewhat divergent, ^{2}H NMR observations by employing deuterated analogs of *n*-alkanes or dimethyl ethylene glycol ethers apparently suggest conformational ordering in conventional nematic liquid crystals such as 4′-methoxybenzylidene-4-*n*-butylaniline and *p*-azoxyanisole [67–79]. When longer-chain molecules are employed as the flexible component, the conventional isotropic phase separation takes place in addition to the order–disorder equilibrium [61–63], yielding a triphasic line at lower temperatures. Flory's lattice model was proved to be useful in reproducing the general feature of such phase diagrams [80]. The conformational aspect of long chains involved in the nematic mesophase is not well understood, however.

Our interest here is to review the thermodynamic character of the flexible chain segment which joins the mesogenic cores on both sides in the so-called mainchain liquid crystals. In the LC state, the flexible segments adjust themselves to make them compatible with the environment. The flexible segment thus takes a mesophase conformation which is different from either the isotropic random-coil or the extended crystalline conformation [26, 81–87].

3.1 Nematic Conformation of Flexible Spacers Involved in the Mesophase

As is known from the pioneering work of Vorlander [88, 89], dimer LC compounds having two terminal mesogens on each end of an intervening spacer often exhibit a very profound odd–even trend in their melting behavior when the melting temperature is plotted against the constituent atoms of the chain. More recently Roviello and Sirigu [90] demonstrated that segmented mainchain polymer liquid crystals exhibit a drastic odd–even oscillation in thermodynamic quantities such as the isotropization temperature and the associated entropy change at the nematic–isotropic transition point. In that study, they also noted that the odd–even fluctuation of the transition entropy is affected by the chemical structure of the functional groups linking the mesogenic unit with the spacer. Such an odd–even effect tends to be largely depressed when carbonate is used in place of the ether or ester groups [90, 91]. Since then a vast number of studies on mainchain polymer liquid crystals have been reported and summarized in various review articles [20, 21].

The molecular weight dependence of the latent entropy $(\Delta S_{\mathrm{NI}})_P$ was studied by Blumstein et al. [92] for a mainchain LC polymer, poly(4,4′-dioxy-2,2′-dimethylazoxybenzene dodecanedioyl) (DDA-9). The fractionated polymer samples, together with the monomer and dimer model compounds were employed in their studies. The value of $(\Delta S_{\mathrm{NI}})_P$ as expressed in terms of a repeating unit increases very rapidly with the degree of polymerization (DP), reaching an asymptotic value in the oligomeric region. When the unit is converted to the entropy change per spacer, the magnitude of $(\Delta S_{\mathrm{NI}})_P$ becomes nearly invariant over a wide range of DP from the dimer (9-DDA-9) to polymers (Fig. 3). These results strongly suggest that the spatial configuration of the flexible spacer and its thermodynamic role remain nearly identical independent of the DP. In general, the DP dependence of physical properties should be most distinct in the oligomer region. The conclusion derived from the observation of Blumstein et al. was later confirmed by comparing the conformational characteristics for a series of mainchain liquid crystals including dimers, trimers, and polymers [93–95].

The ^{2}H NMR technique has provided useful information regarding the orientational characteristics of nematic liquid crystals [96]. In many examples, the order parameters of the mesogenic core comprising a linear array of aromatic nuclei have been accurately determined from the observed dipolar (D_{HD}) and quadrupolar ($\Delta\nu$) splittings by using deuterium-substituted samples. An attempt was made to elucidate the nematic conformation of the polymethylene-type spacer involved in dimer compounds, α,ω-bis(4-cyanobiphenyl-4′-yloxy)alkanes (CBA-n, with $n = 9, 10$), by the combined use of ^{2}H NMR and RIS analysis. The chemical structures and the phase transition data obtained by differential scanning calorimetry (DSC) are shown in

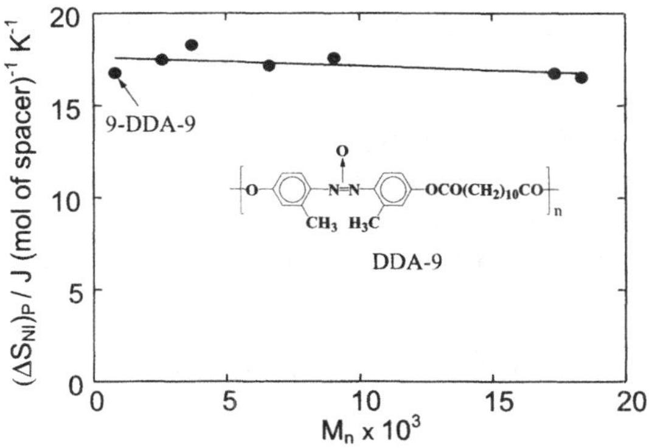

Fig. 3 The isotropization entropy $(\Delta S_{\mathrm{NI}})_P$ vs. the number-average molecular weight M_{n}. The values of $(\Delta S_{\mathrm{NI}})_P$ were those expressed in terms of the numbers of moles of spacers involved. Recalculated from the data (per repeating unit) reported by Blumstein et al. [92]. Note that the dimer model (*9-DDA-9*) involves two mesogenic units and one intervening spacer. The chemical structure of the polymer (*DDA-9*) is shown

Structure 1. Under an NMR magnetic field, the axis of the nematic domain tends to align along the applied field, and thus the resulting LC phase is taken to be of a mono-domain texture.

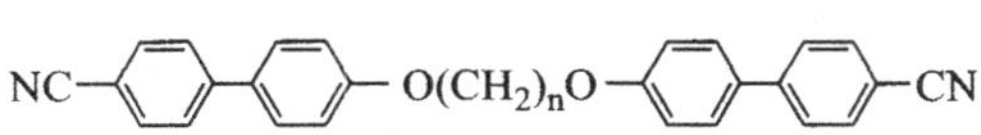

α,ω-Bis(4-cyanobiphenyl-4'-yloxy)alkanes (CBA-n)
n = 9: C 134.8 N 173.0 I (°C)
n = 10: C 164.6 N 184.0 I (°C)

Scheme 1

The procedure, the assumptions underlying the technique, and the results obtained therefrom are briefly summarized as follows:

1. The samples were deuterated at the *ortho* position on both sides of the ether oxygen. The order parameters of the mesogenic core axis S^{R}_{ZZ} were estimated from the observed $D^{\mathrm{R}}_{\mathrm{HD}}$ and $\Delta\nu^{\mathrm{R}}$ according to the prescription described elsewhere [94, 96].

$$D^{\mathrm{R}}_{\mathrm{HD}} = -\gamma_{\mathrm{H}}\gamma_{\mathrm{D}} h / \left(4\pi^2 r^3_{\mathrm{HD}}\right) S^{\mathrm{R}}_{ZZ} , \tag{7}$$

$$\Delta\nu^{\mathrm{R}} = (3/2)\left[S^{\mathrm{R}}_{ZZ} q_{ZZ} + (1/3)\left(S^{\mathrm{R}}_{XX} - S^{\mathrm{R}}_{YY}\right)(q_{XX} - q_{YY})\right] , \tag{8}$$

where γ_{H} and γ_{D} denote the gyromagnetic ratio, h is the Planck constant, r_{HD} is the distance between the deuterium (*ortho*) and proton (*meta*) atoms, and q_{XX}, q_{YY} and q_{ZZ} are the quadrupolar coupling constants defined in the XYZ frame. The numerical values of the parameters required in these expressions may be adopted from the literature [59,97]: $\gamma_H = 2.6752 \times 10^8\ \mathrm{kg^{-1}\,s\,A}$, $\gamma_D = 4.1065 \times 10^7\ \mathrm{kg^{-1}\,s\,A}$, $h = 6.6262 \times 10^{-34}\ \mathrm{J\,s}$, $r_{HD} = 2.48$ Å, $q_{ZZ} = -18.45$ kHz, $q_{XX} = 113.85$ kHz, and $q_{YY} = -95.4$ kHz. The contribution from the biaxiality term $S^R_{XX} - S^R_{YY}$ usually remains small for nematic liquid crystals.

2. The conformation of the spacer was estimated by using samples carrying a perdeuterated polymethylene sequence. The following expression may be adopted for the deuterium quadrupolar splittings ($\Delta\nu_i$) due to the *i*th methylene unit:

$$\Delta\nu_i = (3/2)\left(e^2qQ/h\right) S_{ZZ}\left(3\left\langle\cos^2\phi_i\right\rangle - 1\right)/2\,, \tag{9}$$

where the quadrupolar coupling constant for the aliphatic CD bond is taken to be $(e^2qQ/h) = 174$ kHz [59, 98]. It is assumed here that the molecular axis (z) lies in the direction parallel to the line connecting the centers of the two neighboring mesogenic cores, and that the molecules are approximately axially symmetric around the z-axis in the nematic environment, and thus the orientation of these anisotropic molecules can be described by a single order parameter S_{ZZ}, the biaxiality of the system $S_{XX} - S_{YY}$ being ignored for simplicity. In the equations just given, ϕ_i is the angle between the *i*th CD bond and the molecular axis defined earlier. The bracket indicates statistical mechanical averages taken over all allowed conformations in the system. In this model, the same value of S_{ZZ} is taken to be applicable to all conformers in the mesophase. The order parameter S_{ZZ} may then be approximately related to S^R_{ZZ} of the mesogenic core by

$$S^R_{ZZ} = S_{ZZ}\left(3\left\langle\cos^2\psi\right\rangle - 1\right)/2\,, \tag{10}$$

where ψ denotes the disorientation of the mesogenic core axis with respect to the molecular axis. The analysis of the experimental results may be facilitated by taking ratios such as

$$\Delta\nu_i/\Delta\nu^R = \mathrm{const}\left(3\left\langle\cos^2\phi_i\right\rangle - 1\right)/\left(3\left\langle\cos^2\psi\right\rangle - 1\right)\,, \tag{11}$$

$$\Delta\nu_i/\Delta\nu_j = \mathrm{const}\left(3\left\langle\cos^2\phi_i\right\rangle - 1\right)/\left(3\left\langle\cos^2\phi_j\right\rangle - 1\right) \quad (i \neq j)\,. \tag{12}$$

These ratios should solely depend on the spacer conformation, being free from the orientational order of the molecular axis S_{ZZ}. The values of $\Delta\nu_i/\Delta\nu^R$ as well as $\Delta\nu_i/\Delta\nu_j$ have been found to remain nearly invariant with temperature, suggesting that the conformational correlation along the spacer is governed by the same principle throughout the nematic domain. Since the flexible spacer maintains liquidlike characteristics in the nematic state, the equilibration among the conformers should be affected by temperature, leading to a small variation in the ratio if the data are collected over a wider temperature range.

3. The mean-square average $\langle\cos^2\phi\rangle$ for given CD bonds may be estimated by using the conformer distribution derived from the RIS analysis. In

practice the nematic ensemble of the chain configuration could be elucidated by performing an iterative RIS calculation on the basis of the conformer distribution map until the experimentally observed quadrupolar splitting ratios (Eqs. 11, 12) are satisfactorily reproduced. With the configurational partition function Z_N thus estimated, the conformational entropy change S_{NI}^{conf} at the NI interphase may be elucidated by

$$S_{NI}^{conf} = k \ln(\widetilde{Z}) + kT \; d\ln(\widetilde{Z})/dT\,, \tag{13}$$

where $\widetilde{Z} = Z_I/Z_N$. Likewise, the S_{CN}^{conf} corresponding to the CN transition may be obtained by setting $Z = Z_N$ in Eq. 1. (Note that $Z = 1$ for the crystalline state.)

Extensive structure analyses were first carried out for the dimers depicted in Scheme 1 (CBA-n, $n = 9, 10$). The same techniques were applied to the main-chain polymer liquid crystals having analogous chain sequences (Scheme 2), leading to the conclusion that conformational characteristics of the spacer involved in the nematic state are essentially the same in both dimer and polymer liquid crystals [99]. The molecular scheme described previously has also been successfully adopted in the analysis of the observed ^{13}C chemical shift and the ^{13}C–^{13}C dipolar coupling data of the dimer and trimer compounds [100, 101].

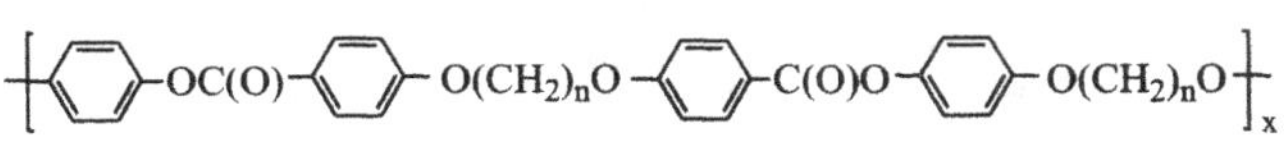

Mainchain polymer LCs with dialkoxy spacers

n = 9: C 140 N 200 I (°C)
n = 10: C 169 N 204 I (°C)

Scheme 2

In view of the unique character of the carbonate compounds [90], the ^{2}H NMR RIS treatment was performed on the nematic dimer liquid crystals comprising carbonate linkages (Scheme 3) [91]:

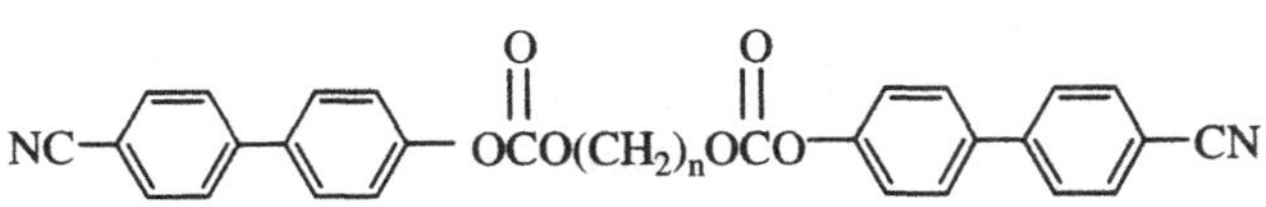

α,ω-Bis(4-cyanobiphenyl-4'-yloxycarbonyloxy)alkanes (CBC-n, n = 5,6)

n = 5: C 188.0 N 192.5 I (°C)
n = 6: C 183.0 N 189.0 I (°C)

Scheme 3

The average conformation was estimated for each of the constituent bonds of the spacer. In Fig. 4, the variation of the *trans* fraction (f_t) is plotted along the spacer from the terminal to the central bond of the ether dimer CBA-n ($n = 9, 10$) (open circles) and the carbonate dimer CBC-n ($n = 5, 6$) (filled circles). For illustrative purposes, the n is odd (Fig. 4a) and the n is even (Fig. 4b) series are shown separately. The odd–even oscillation of the bond conformation is indicative of a long-range intramolecular correlation along the spacer. The bond conformations calculated at various temperatures were found to be nearly identical within the nematic state. The suppression of unfavorable molecular orientations in the nematic constraint leads to a characteristic conformational ordering regardless of the chemical structure of the linking groups. The effect of the linking group is manifestly shown in the oscillation of the f_t values for the first three bonds. As a consequence, the odd–even trend of the bond conformation becomes similar between CBC-5 and CBC-6 in conformity with the weak odd–even character known for various thermodynamic quantities of these compounds. In a separate experiment on CBC-n, the order parameter of the mesogenic cores obtained in the neighborhood of the NI transition point exhibits only weak odd–even alternation

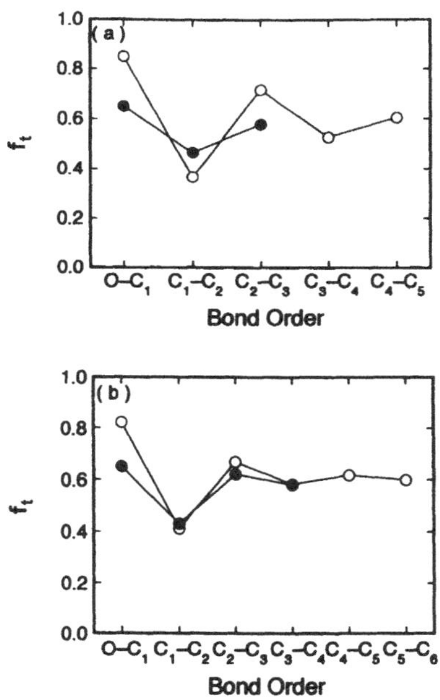

Fig. 4 Bond conformation of the internal O – C and C – C bonds of the spacer in the nematic phase. The fractions of the *trans* conformer (f_t) are shown in the order from the terminal to the central bond. The values of f_t for the carbonate-type dimer liquid crystals are compared with those of the ether-type: (**a**) CBC-5 (•) and CBA-9(○),(**b**) CBC-6 (•) and CBA-10 (○). (Reproduced from Ref. [91] with permission)

with n. These observations indicate that the flexible spacer takes a conformation characterized by a given ensemble of spatial arrangements in the LC state for both CBA-n and CBC-n. The results of the conformational analysis described earlier lead to the conclusion that the origin of the odd–even effect can be traced back to the relative inclination of the mesogenic core axes joined by an intervening spacer, which is very sensitive to the bond angle of the linking groups [6, 91]. The effect due to the difference in the rotational characteristics around the internal bonds seems to be secondary.

The analysis was further extended to include trimer compounds, 4,4′-bis[ω-(4-cyanobiphenyl-4′-yloxy)alkoxy]biphenyls (CBA-Tn, with $n = 9, 10$) (Scheme 4), comprising three mesogenic units joined by spacers [95, 100, 101], and several mainchain compounds having oxyethylene-type spacers, designated as BuBE-x (Scheme 5), and MBBE-x (Scheme 6) [102, 103]. In the last example, designated as MBBE-6, the contour length of the spacer $-(OCH_2CH_2)_6-$ (21.4 Å) exceeds that of the hard segments (18.2 Å).

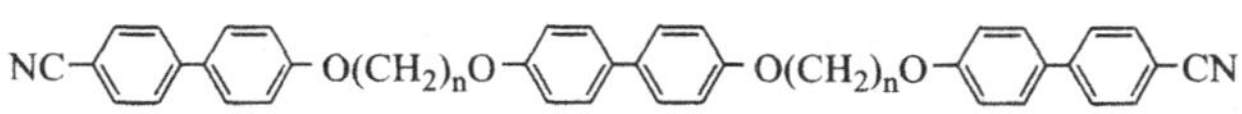

4,4'-Bis[ω-(4-cyanobiphenyl-4'-yloxy)alkoxy]biphenyls (CBA-Tn)

n = 9: C 146.0 N 195.7 I (°C)
n = 10: C 188.5 N 209.9 I (°C)

Scheme 4

BuBE-x

x = 2: C 124.6 N 132.2 I (°C)
x = 3: C 99.0 N 103.0 I (°C)

Scheme 5

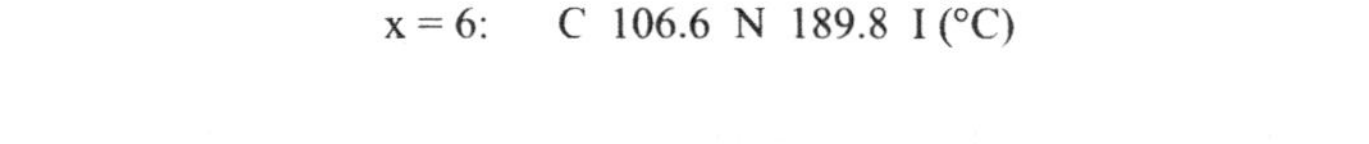

MBBE-x

x = 2: C 191.8 N 301.5 I (°C)
x = 3: C 165.4 N 283.4 I (°C)
x = 4: C 148.0 N 242.9 I (°C)
x = 5: C 132.2 N 216.1 I (°C)
x = 6: C 106.6 N 189.8 I (°C)

Scheme 6

In all these examples, the quadrupolar splitting ratios such as $\Delta\nu_i/\Delta\nu^{R}$ and $\Delta\nu_i/\Delta\nu_j$ $(i \neq j)$ were found to remain nearly invariant over a wide range of temperature, which is indicative of the nematic conformation. Some of these liquid-crystal-forming dimer and trimer compounds were adopted in the following *PVT* studies as a model for the mainchain-type polymer liquid crystals on the assumption that the spatial configuration and thus the thermodynamic roles of the flexible spacer are nearly identical as long as the chemical structure of the repeating unit is similar. It is advantageous to work with oligomeric model compounds having neat chemical structures. Shortcomings inherent to polymeric liquid crystals such as polydispersity in the DP, imperfections due to the irregular arrangements (e.g., kink conformation [22, 23, 104]), and difficulties associated with higher NI transition temperatures can thus be avoided.

The analysis of binary mixtures with a low molar mass (monomer) liquid crystal gave important information regarding the mesophase structure of CBA-*n* [105] as well as CBA-T*n* [95]. The ordering characteristics of the two components of the mixture can be elucidated by comparing the orientational order parameters estimated independently. In one of the examples where trimer CBA-T9 was dissolved in a monomer liquid crystal, 4′-ethoxybenzilidene-4-cyanoaniline, the order parameter of the monomer was found to be significantly higher than that (S^{R}_{ZZ}) of the trimer in the nematic mixture at any given temperature and concentration, indicating that the spatial arrangements of the mesogens are largely restricted in the trimer owing to the stereochemical requirement of the intervening spacer. The schematic diagram shown in Fig. 5 indicates the nematic arrangements of the ensemble thus deduced and the transitions to the adjacent isotropic and crystalline phases.

The combined use of spectroscopic and thermodynamic techniques on CBA-*n* has led to a general picture for the phase transition: Eq. 1 Under a uniaxial potential field, both spacer and mesogenic units at the terminals tend to align along the domain axis. Consequently, the individual mesogenic cores

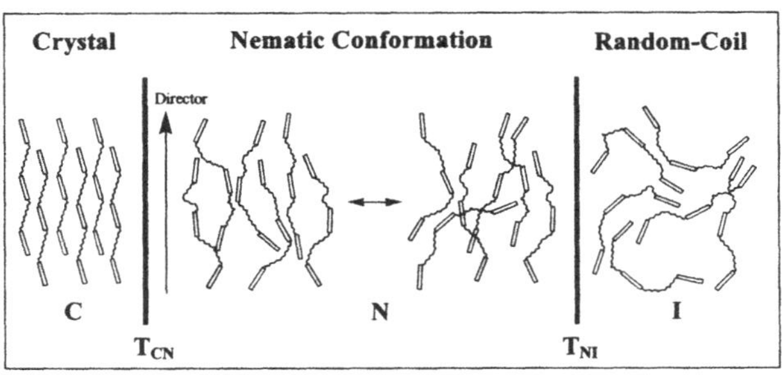

Fig. 5 Schematic representation of the crystal–nematic (CN) and nematic–isotropic (NI) transitions, illustrated for a trimer liquid-crystal system. The conformational distribution of the spacer remains quite stable in the nematic phase, and is termed *nematic conformation*

inevitably incline to some extent with respect to the direction of the molecular extension (Fig. 5), giving rise to a moderate value of the orientational order parameter (S^{R}_{ZZ}) (Eq. 2). As revealed by studies of magnetic susceptibility [106, 107] as well as optical anisotropy [108], the molecular anisotropy of dimer compounds CBA-n ($n = 9, 10$) increases on going from the isotropic to the nematic LC state. Although the flexible spacer takes a more extended conformation in the LC state, the contribution of the spacer to the attractive part of the orientation-dependent intermolecular interactions seems to be small (Eq. 3). While the orientational fluctuation of the entire molecule varies as a function of temperature, the nematic conformation of the spacer remains nearly invariant over the entire range of the LC state (Eq. 4). It has been confirmed by ^{2}H NMR and *PVT* studies that 50–60% of the transition entropy $(\Delta S_{\mathrm{tr}})_P$ arises from the variation in the conformational distribution of the spacer at the phase boundary [93–95]. It should be noted here that the order–disorder characteristics inherent in the primary structure of the flexible spacer are precisely controlled in order to develop the LC mesophase.

The conclusions derived from the polymethylene-type liquid crystals are safely applicable to those comprising the oxyethylene-type spacers. The formation of the nematic conformation characteristic of the LC state has been confirmed using the ^{2}H NMR technique on deuterated samples of BuBE-x [102] as well as MBBE-x [103]. In these compounds, the $-\mathrm{OC}-\mathrm{CO}-$ bond tends to take a *gauche* form in the fluid state in contrast to the all-*trans* preference of the polymethylene-type spacer. The results of the IR analysis indicate that the *gauche* forms are highly populated in the LC phase in all samples examined [109]. When chains are placed in a uniaxial potential field such as a nematic environment, the correlation of bond rotations along the chain should become important. The general rule underlying such rotational correlations is not obvious for the oxyethylene-type spacer. A precise determination of the nematic conformation of this type of spacer is an interesting subject, which has not been completed.

3.2
PVT Analysis of Segmented Compounds Capable of Forming Liquid Crystals

In the nematic state, the molecular orientations are governed by a uniaxial potential field, and thus the spacers are required to recognize a certain conformational correlation along the chain (nematic conformation), but the overall properties are still very much liquidlike. The liquid–liquid first-order transition between the isotropic and nematic states presents an interesting example to investigate the validity of the entropy separation hypothesis (see Sect. 2.2).

Most polymer samples exhibit the NI transition well beyond 200 °C [43, 110, 111]. Since thermal stability is important for these measurements, the *PVT* data were mainly collected for the dimer and trimer compounds listed in the preceding section. While the depression of the phase transition temperature due to the supercooling effect, $\Delta T = T_{tr}(\text{heating}) - T_{tr}(\text{cooling})$, is found to be marked for T_{CN}, the corresponding difference is almost nil for T_{NI}. The data obtained from the experiments under the isothermal condition during the heating cycle were treated to elucidate the volume change ΔV_{tr} at the transition temperature, and the γ vs. V_{sp} relation was constructed [43].

3.2.1 Determination of the Thermal Pressure Coefficient of Mainchain LC Compounds

A typical example of the isothermal γ vs. V_{sp} curves is given in Fig. 6 for the dimer compound designated as MBBE-6 (DSC data are given in Sect. 3.1) [109]. The γ values at the transition temperature were estimated by extrapolation from those obtained in the stable region. The volume changes, $\Delta V_{NI}(0.0042\ \text{cm}^3\ \text{g}^{-1})$ and $\Delta V_{CN}(0.0485\ \text{cm}^3\ \text{g}^{-1})$, taking place at the NI and CN transitions are indicated by the dotted lines. MBBE-6 has a relatively long spacer, exhibiting a nematic LC phase over a temperature range 107–190 °C. The *PVT* measurements carried out in high precision yielded a quite good estimate for all three states, i.e., crystal, nematic liquid crystal, and isotropic melt. Figure 6 may be used to elucidate γ values according to both methods 1 and 2 prescribed in Sect. 2.2.2. The $\gamma - V_{sp}$ curves for the transition temperature can be extrapolated (if necessary) to the phase boundary of the biphasic gap. In the diagram, the γ_{tr} values of method 1 can be defined at

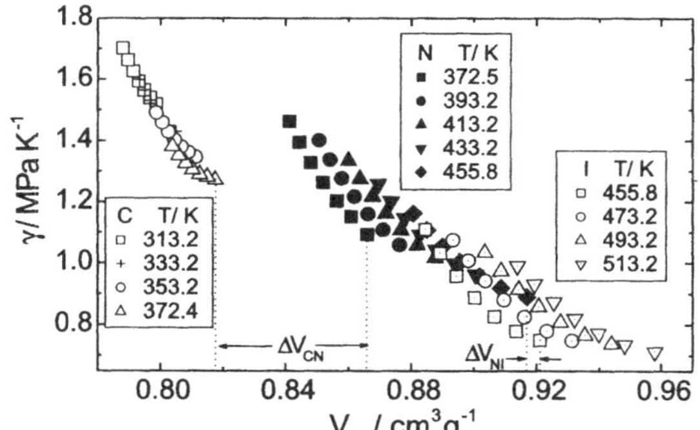

Fig. 6 The γ vs. V_{sp} relation of MBBE-6 estimated for given temperatures in each phase. The γ values corresponding to the transition temperatures ($T_{NI} = 455.8$ K, $T_{CN} = 372.4$ K) were estimated by extrapolation. The incidences of the phase transition (ΔV_{CN} and ΔV_{NI}) are indicated by the *dotted lines*

the point where the curve intersects the dotted line (the phase boundary): $\gamma_{NI}(I) = 0.75$ and $\gamma_{NI}(N) = 0.89$ MPa K^{-1} (at $T_{NI} = 455.8$ K), and $\gamma_{CN}(N) = 1.09$ and $\gamma_{CN}(C) 1.27$ MPa K^{-1} (at $T_{CN} = 372.4$ K). Here the related phases are indicated in parentheses. The same $\gamma - V_{sp}$ curves can be further extrapolated to the midpoint of the biphasic gap (ΔV_{NI} and ΔV_{CN}) to get the γ_{tr} values of method 2: $\gamma_{NI}(I) = 0.76$ and $\gamma_{NI}(N) = 0.88$ MPa K^{-1} for the NI transition, and $\gamma_{CN}(N) = 1.44$ and $\gamma_{CN}(C) = 1.25$ MPa K^{-1} for the CN transition. The disparities between the two sets of γ values obtained with method 1 are as follows: $\gamma_{NI}(I)/\gamma_{NI}(N) = 0.84$ and $\gamma_{CN}(N)/\gamma_{CN}(C) = 0.86$. The corresponding ratios for method 2 are $\gamma_{NI}(I)/\gamma_{NI}(N) = 0.86$ and $\gamma_{CN}(N)/\gamma_{CN}(C) = 1.15$. Although the discrepancies between a pair of measurements are definite for both NI and CN transitions, the amount (approximately 15%) is reasonably small for the purpose at hand. As will be shown later, the choice of γ either from method 1 or from method 2 is rather unimportant in terms of the entropy changes ΔS_V as well as $(\Delta S_{tr})_V$. In conformity with the treatment of the other dimer and trimer compounds, the γ values obtained according to method 1 are listed in Tables 2 and 3.

The values of γ were similarly obtained for dimer CBA-n ($n = 9, 10$) and trimer CBA-Tn ($n = 9, 10$). These compounds exhibit the nematic LC phase over a limited temperature range, hampering an accurate estimation of γ by the extrapolation from this phase. Accordingly the γ values were estimated by method 1 only from higher-temperature phases: i.e., γ_{NI} values are estimated from the isotropic phase, and γ_{CN} values from the nematic phase [95]. The γ_{tr} values thus derived are all accommodated in Tables 2 and 3, respectively, for the NI and CN transitions. Thermal pressure coefficients of monomer liquid crystals such as 4-cyano-4′-alkylbiphenyls (nCB) and 4-cyano-4′-alkoxybiphenyls (nOCB) are available in the article by Orwoll et al. [112]. The γ values applicable to the NI transition of these compounds are cited in Table 4 for comparison. As shown in these tables, use of the volume change ΔV_{tr} at the transition (column 4) leads to the estimate of the volume-dependent entropy ΔS_V (column 5) according to Eq. 3.

3.2.2 Estimation of Constant-Volume Transition Entropies at the NI and CN Interphase

The transition entropies under atmospheric pressure can be conventionally determined by DSC. The heats estimated by DSC are often affected by the amount of sample loaded on the pan and also on the scanning speed of the apparatus. They may include some additional contributions due to pretransitional and posttransitional phenomena occurring in the immediate vicinity of the transition [113–117]. In addition, some difficulty is inevitably invoked in selecting a baseline to measure the area of the absorption peak. In view of these difficulties involved in the DSC method, the use of the Clapeyron rela-

Table 2 Thermal pressure coefficients γ, volume changes ΔV, and volume-dependent transition entropies ΔS_V of dimer and trimer liquid crystals (*LCs*) for the nematic–isotropic (NI) phase transitions

Compound	T_{NI} (K)	γ (MPa K^{-1})	ΔV_{NI} (cm^3 mol^{-1})	NI transition ΔS_V (= $\gamma \Delta V_{NI}$) (J mol^{-1} K^{-1})	$(\Delta S_{NI})_P$ (J mol^{-1} K^{-1})	$(\Delta S_{NI})_V$ [= $(\Delta S_{NI})_P - \Delta S_V$] (J mol^{-1} K^{-1})	S^{conf} (J mol^{-1} K^{-1})
				Dimer LCs with polymethylene-type spacers			
CBA-9	468.9[a]	0.93[a]	7.5[a]	7.0	14.9[a], 15.3[b], 13.1[c]	7.9[a], 8.3[b], 6.1[c]	13.3[d]
CBA-10	466.9[a]	0.94[a]	9.4[a]	8.8	22.1[a], 15.3[b], 19.0[c]	13.3[a], 6.5[b], 10.2[c]	15.6[d]
				Trimer LCs with polymethylene-type spacers			
CBA-T9	468.9[e]	0.88[e]	11.7[e]	10.3	25.0[e], 20.8[c]	14.7[e], 10.5[c]	(26.6)[f]
CBA-T10	483.1[e]	0.97[e]	19.9[e]	19.3	49.8[e], 43.2[c]	30.5[e], 23.9[c]	(31.2)[f]
				Dimer LC with an oxyethylene-type spacer			
MBBE-6	455.7[g]	0.75 (I)[g]	4.0	3.0 (I)[g]	8.98[g] 7.04[h]	5.98(I)[g], 4.04 (I)[h]	
		0.89 (N)[g]		3.56 (N)[g]		5.42 (N)[g], 3.48 (N)[h]	

[a] Ref. [93]
[b] Ref. [118]
[c] Ref. [119]
[d] Ref. [94]
[e] Ref. [95]
[f] The values are twice those for the corresponding dimers (cf. the preceding rows).
[g] Ref. [109]: those derived from the γ value estimated by extrapolation from the isotropic, nematic, and solid phases are, respectively, distinguished by I, N, and C
[h] Ref [109]: estimated by differential thermal analysis

Table 3 Thermal pressure coefficients γ, volume changes ΔV, and volume-dependent transition entropies ΔS_V of dimer and trimer LCs for the crystal–nematic (CN) phase transitions

Compound	T_{CN} (K)	γ (MPa K^{-1})	ΔV_{CN} (cm^3 mol^{-1})	CN transition ΔS_V (= $\gamma \Delta V_{CN}$) (J mol^{-1} K^{-1})	$(\Delta S_{CN})_P$ (J mol^{-1} K^{-1})	$(\Delta S_{CN})_V$ [= $(\Delta S_{CN})_P - \Delta S_V$] (J mol^{-1} K^{-1})	S^{conf} (J mol^{-1} K^{-1})
				Dimer LCs with polymethylene-type spacers			
CBA-9	412.2[a]	1.76[a]	37.5[a]	66.0	147.0[b], 127.1[c]	81.0[b], 61.1[c]	59.6[d]
CBA-10	438.2[a]	1.72[a]	39.3[a]	67.4	143.8[b], 132.0[c]	76.4[b], 64.6[c]	64.2[d]
				Trimer LCs with polymethylene-type spacers			
CBA-T9	419.2[e]	1.67[e]	58.1[e]	97.0	222.5[c]	125.5[c]	(119.2)[f]
CBA-T10	461.7[e]	1.32[e]	78.0[e]	103.1	226.0[c]	122.9[c]	(128.4)[f]
				Dimer LC with an oxyethylene-type spacer			
MBBE-6	372.4[g]	1.09 (N)[g]	45.8[g]	49.9 (N)[g]	158.9[g], 120.9[h]	109.0 (N)[g], 71.0 (N)[h]	
		1.27 (C)[g]		58.2 (C)[g]		100.7 (C)[g], 62.7 (C)[h]	

[a] Ref. [93]
[b] Ref. [118]
[c] Ref. [119]
[d] Ref. [94]
[e] Ref. [95]
[f] The values are twice those for the corresponding dimers (cf. the preceding rows).
[g] Ref. [109]: those derived from the γ value estimated by extrapolation from the isotropic, nematic, and solid phases are, respectively, distinguished by I, N, and C
[h] Ref [109]: estimated by differential thermal analysis

Table 4 Thermal pressure coefficients γ, volume changes ΔV, and volume-dependent transition entropies ΔS_V of monomer LCs for the NI phase transitions [112]

Compound	T_{NI} (K)	γ (MPa K^{-1})	ΔV_{NI} (cm^3 mol^{-1})	ΔS_V (J mol^{-1} K^{-1})	$(\Delta S_{NI})_P$ (J mol^{-1} K^{-1})	$(\Delta S_{NI})_V$ (J mol^{-1} K^{-1})
5CB	308.4	1.46	1.06	1.5	2.1	0.6
7CB	315.9	1.46	1.62	2.4	2.8	0.4
5OCB	340.6	1.39	0.78	1.1	1.3	0.2
8OCB	352.9	1.16	1.33	1.5	2.3	0.8

tion, which provides a measure of the discontinuity of the first-order phase transition at a given temperature, is more preferable.

$$\Delta S_{tr} = \Delta V_{tr}\, dp/dt\,. \tag{14}$$

The slope of the phase boundary curve dp/dt can be estimated from the $T_{tr} - P_{tr}$ plot obtained by using a *PVT* or high-pressure differential thermal analysis (DTA) method [118, 119]. The volume-dependent entropy (correction for the volume change) ΔS_V, the transition entropy $(\Delta S_{tr})_P$ under ordinary pressure, and the constant-volume entropy $(\Delta S_{tr})_V$ obtained therefrom are arranged in this order in Tables 2, 3, and 4.

As comparison of Tables 2 and 4 indicates, the changes in volume and entropy at the NI transition obtained for the mainchain dimer and trimer liquid crystals are much larger than those reported for conventional monomer liquid crystals [112]. In Tables 2 and 3, the constant-volume transition entropies $(\Delta S_{tr})_V$ are expressed in terms of joules per mole per kelvin. The conformational entropy changes S_{tr}^{conf} estimated on the basis of the ^{2}H NMR quadrupolar splitting data observed in the LC state are as follows: $S_{NI}^{conf} = 13.3$ (CBA-9) and 15.6 (CBA-10), and $S_{CN}^{conf} = 59.6$ (CBA-9) and 64.2 (CBA-10), the units being joules per mole per kelvin [94]. While the values of S_{NI}^{conf} somewhat exceed those of $(\Delta S_{NI})_V$ from *PVT*, S_{NI}^{conf} tends to be slightly lower than $(\Delta S_{CN})_V$ (columns 7 and 8). In view of the long derivation required in these estimations, the correspondence between the two numerical figures is reasonable for both transitions of CBA-*n*. Since the trimer compounds comprise two spacers, values twice those of the conformational entropies S_{tr}^{conf} given previously for the dimer are shown in parentheses in the last columns of Tables 2 and 3. Comparison with the *PVT* values of $(\Delta S_{NI})_V$ exhibits some deviation for the NI transition. It may also be interesting to examine the ratio such as $(\Delta S_{tr})_V$(CBA-T*n*) vs. $(\Delta S_{tr})_V$(CBA-*n*), which measures the effect due to the number of spacers involved. For the NI transition, the ratios are 1.5 ($n = 9$) and 2.8 ($n = 10$), and for the CN transition, 1.8 ($n = 9$) and 1.7 ($n = 10$). If these entropies are exactly additive in terms of the spacer, the ratio is supposed to be 2. In general, the aforementioned description of the nematic conformation (Fig. 5) is reasonably consistent with the *PVT* observations. The present analysis supports the view that most of the constant-volume transition entropy $(\Delta S_{NI})_V$ should arise from the difference in the conformer distribution between the two phases in equilibrium.

For MBBE-6, the γ_{tr} values are also available from method 2 (Sect. 3.2.1). As mentioned in the preceding section, the difference between methods 1 and 2 is quite small except for $\gamma_{CN}(N)$. An increment of 25% in $\gamma_{CN}(N)$ gives rise to a decrease in $(\Delta S_{CN})_V$ by approximately 20% from those listed in column 7 of Table 3. As shown in Tables 2 and 3, the experimental values of the latent entropy $(\Delta S_{tr})_P$ (column 6) are quite divergent depending on the method (*PVT* or DTA) employed in the measurement. This is also a source of the uncertainties in the final estimates of the constant-volume transition entropy $(\Delta S_{tr})_V$ (column 7).

At the NI transition, an orientation-dependent term such as ΔS^{orient} must play a role [26]. In practice, however, the contribution from this source seems to be comparatively small by inference from those of the monomer liquid crystals [112] (cf. Table 4). In treating the transition entropy of real systems, contributions from the so-called communal entropy as well as other residual entropies are often considered by introducing an extra term ΔS_d in Eq. 1 [15, 17, 34, 35, 120]:

$$(\Delta S_{tr})_V = (\Delta S_{tr})_P - \Delta S_V - \Delta S_d \, . \tag{15}$$

The physical definition of the ΔS_d term is still obscure for polymeric systems in which the external degrees of freedom are largely restricted by the chain connectivity. In general, the contribution from this source to the total entropy change is assumed to be small and is often ignored for polymeric chains [120]. As stated already, the nematic phase is very much liquidlike, and the component molecules still maintain their translational freedom. The contribution from the residual entropies ΔS_d must be much smaller than those involved in the phase transition between the crystal and the isotropic melt.

4
Concluding Remarks

In this article, the conformational entropy changes taking place during the crystallization/melting via a nematic mesophase were the major concern. The nematic conformation is a newly found form of chain molecules, which is different from those of the neighboring phases. Dimer and trimer model compounds were mostly treated on the assumption that the thermodynamic role of the spacer in determining the phase transition is similar to those of polymer liquid crystals having the same chain segments. We have not discussed the morphological effect of the interstitial nematic mesophase on the crystallization.

In Fig. 7, X-ray profiles are shown for two MBBE-6 samples prepared in two ways: melt-crystallized and solution-grown [103]. While the former was prepared by crystallization from the melt through the intervening nematic phase, the latter was obtained by precipitation from a chloroform solution with addition of methanol. The two diagrams are nearly identical for annealing at a temperature 10 degrees below the melting point. The uniaxial alignment maintained in the LC phase was destroyed during the nucleation-growth process of the crystallization. Crystallization under a magnetic field (e.g., in an NMR tube) yielded partial orientation of crystallites, indicating that the nematic alignment of chain segments precipitates some residual effect on the orientation of the crystals formed. The same behavior is commonly observed for all dimer and trimer samples. Polymeric liquid crystals must largely differ in their dynamic properties The results of neutron diffraction studies on nematic liquid crystals indicate that the long polymer chains are confined in a long cylinder with few hairpin defects to fold back [22, 23, 121]. In most examples, mainchain liquid crystals eventually crystallize at lower temperatures. Whether the formation of a stable mesophase prior to crystallization enhances the following nucleation step is not well understood, however. After Kaji et al. [122], a substantial number of observations have been accumulated in favor of the "mesomorphic phase" model as a mechanism of the cold-crystallization of polymers [123]. In the crystallization of chain molecules, both density and orientation should play an important role. The stepwise crystallization of chain molecules via a nematic mesophase may provide some valuable information regarding the embryonic stage of the polymer crystallization mechanism [124].

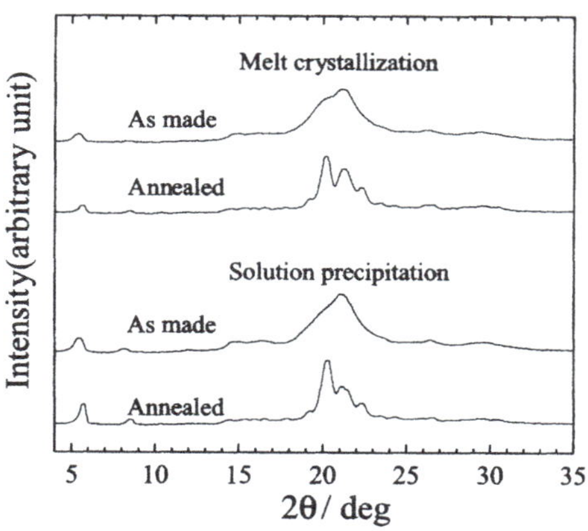

Fig. 7 X-ray diffraction profiles of two MBBE-6 samples prepared by the melt-crystallization and solution-precipitation processes. The two diagrams became nearly identical for annealing at a temperature 10 degrees below the melting point (379.8 K)

On going from the isotropic to the anisotropic LC state, the orientation-dependent attractive interactions come into play [125, 126] while the steric interactions (the excluded-volume effect) between the mesogenic rods are relaxed [127]. In the LC state, all molecules are required to take an asymmetric shape. Accordingly, chain segments adopt a unique conformer distribution called a nematic conformation [26, 93–95, 102, 103, 105]. The $\gamma - V_{sp}$ relation determined in the aforementioned studies may be used to examine the mean-field potentials effective in nematic as well as isotropic liquids. After Frank [128] and Hildebrand and Scott [2, 129], the intermolecular interaction potentials such as

$$\varepsilon = -\eta/V^m \tag{16}$$

or somewhat modified forms have been widely adopted in conventional mean-field theories of liquids including molten (amorphous) polymers [130–133]. Here η is defined as a mean-field parameter representing the strength of the interaction field and the parameter m is empirically taken to be a constant in the range 1–2. Thus,

$$T\gamma = (\partial E/\partial V)_T = -m\eta/V^{m+1} \,. \tag{17}$$

Assuming that η is independent of volume and temperature,

$$\ln T\gamma = -(m+1)\ln V + \ln m\eta \,. \tag{18}$$

A plot of $\ln T\gamma$ vs. $\ln V_{sp}$ should give an estimate of m and η, respectively, from the slope and the intercept of the curve. Figure 8a indicates the variation of m as a function of temperature at zero pressure for MBBE-6 [45]: m approaches unity (dotted line) only within a narrow range of temperature. The analysis suggests that the conventional form of the potential given in Eq. 16 is only valid within a certain limited range of temperature and volume in both isotropic and nematic LC phases. In the conventional treatment of liquids, m is often set equal to unity in Eq. 16 [2, 129]. The $\eta - T$ curves derived for $m = 1$ are illustrated in Fig. 8b. When the reduced temperature scale such as $T^* = T/T_{tr}(T_{tr} = T_{CN}$ or $T_{NI})$ is employed in each phase, the two separate curves in Fig. 8, one for the isotropic and the other for the nematic phase, come closer in both $m - T$ and $\eta - T$ plots. In this connection, it might be interesting to note that the thermal expansion coefficient α vs. T plot exhibits a similar trend. According to the equation of state of liquids prescribed by Flory [131, 134],

$$(V/V^*)^{1/3} - 1 = \alpha T/3(1 + \alpha T) \,, \tag{19}$$

where V^* designates the core volume. The free volume $f_V = (V - V^*)/V$ of the isotropic melt estimated around the vicinity of the NI transition amounts to approximately 0.2 for all dimer and trimer compounds. Since the thermal expansion coefficient does not change much on going from the isotropic melt to the nematic LC state, the free volume should remain nearly the same between

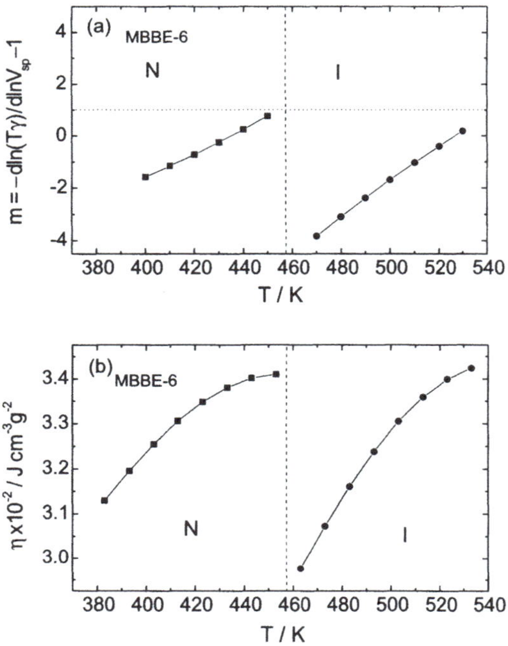

Fig. 8 Variation of the intermolecular interaction parameters (**a**) m obtained directly from Eq. 18 and (**b**) η (estimated with $m = 1$ in Eq. 18) as a function of temperature at zero pressure for MBBE-6. The NI phase transition temperature is indicated by the *broken lines*. In **a**, the value of m approaches unity (*dotted line*) only within a narrow range of temperature in both phases. The abnormal behavior observed in the vicinity of the transition point [113–117] is not shown

these two states. The dimer or trimer liquid crystal molecules are highly mobile in the nematic state. A further analysis of the *PVT* data should yield more detailed knowledge about the characteristic feature of the intermolecular interactions of chain molecules accommodated in the nematic state.

Finally we wish to emphasize that accumulation of reliable experimental data on the conformational ordering and thermodynamic characteristics is a prerequisite for the establishment of molecular theories of liquid crystals consisting of two components of very different flexibilities [80, 135–137].

Acknowledgements This work was partially supported by the Private University Frontier Research Center Program sponsored by the Ministry of Education, Culture, Sports, Science, and Technology (Monbukagakusho), and also by a Grant-in-Aid for Science Research (15550111) of Monbukagakusho. One of the authors (Z.Z.) would like to express his gratitude for the Postdoctoral Fellowship for Foreign Researchers sponsored by the Japan Society for the Promotion of Science, and he is also thankful for the support by a Grant-in-Aid for Science Research (P01094) of Monbukagakusho.

References

1. NIST standard reference data program:chemistry webbook: http://webbook.nist.gov/ National Institute of Standards and Technology, Gaithersburg, MD, USA
2. Hildebrand JH, Scott RL (1950) The solubility of nonelectrolytes. ACS monograph. Reinhold, New York
3. Flory PJ (1953) Principles of polymer chemistry, Cornell University Press, New York
4. Mandelkern L (1964) Crystallization of polymers. McGraw-Hill, New York
5. Wunderlich B, Czornyj G (1977) Macromolecules 10:906
6. Flory PJ (1969) Statistical mechanics of chain molecules. Wiley-Interscience, New York
7. Mattice WL, Suter UW (1994) Conformational theory of large molecules. Wiley, New York
8. Starkweather HW, Boyd RH (1960) J Phys Chem 64:410
9. Kirshenbaum I (1965) J Polym Sci A 3:1869
10. Smith RP (1966) J Polym Sci A-2 4:869
11. Tonelli AE, Srinivasarao M (2001) Polymers from the inside out. Wiley-Interscience, New York
12. Tsujita Y, Nose T, Hata T (1972) Polym J 3:587
13. Tsujita Y, Nose T, Hata T (1974) Polym J 6:51
14. Sundararajan PR (1978) J Appl Polym Sci 22:1391
15. Naoki M, Tomomatsu T (1980) Macromolecules 13:322
16. Abe A (1980) Macromolecules 13:546
17. Wurflinger A (1984) Colloid Polym Sci 262:115
18. Wunderlich B (1980) Macromolecular physics, vol 3. Crystal melting. Academic, New York
19. Mandelkern L (2002) Crystallization of polymers, 2nd edn. Cambridge University Press, Cambridge
20. Blumstein A, Asrar J, Blumstein RB (1984) In: Griffin AC, Johnson JF (eds) Liquid crystals and ordered fluids, vol 4. Plenum, New York, p 311
21. Sirigu A (1991) In: Ciferri A (ed) Liquid crystallinity in polymers. VCH, New York, p 261
22. Davidson P (1999) In: Mingos DMP (ed) Liquid crystals II. Springer, Berlin Heidelberg New York,p 1
23. Cotton JP, Hardouin F (1997) Prog Polym Sci 22:795
24. Ellis G, Gomez M, Marco C (2000) Analysis 28:22
25. Boeffel C, Spiess HW (1997) In: Chiellini E, Giordano M, Leporini D (eds) Structure and transport properties in organized polymeric materials. World Scientific, Singapore, p 125
26. Abe A (1984) Macromolecules 17:2280
27. Imrie CT (1999) In: Mingos DMP (ed) Liquid crystals II. Springer, Berlin Heidelberg New York, p 149
28. Abe A, Furuya H (1988) Polym Bull 19:403
29. Abe A, Furuya H (1988) Polym Bull 20:467
30. Abe A, Furuya H (1988) Macromolecules 22:2982
31. Mark JE (1977) J Chem Phys 67:3300
32. Abe A (1989) In: Booth C, Price C (eds) Comprehensive polymer science, vol 2. Pergamon, Oxford, p 49
33. Flory PJ (1976) J Macromol Sci Phys Ed B 12:1
34. Turturro A, Bianchi U (1975) J Chem Phys 62:1668

35. Bianchi U, Turturro A (1976) J Chem Phys 65:697
36. Karasz FE, Couchman PR, Klempner D (1977) Macromolecules 10:88
37. Bleha T (1985) Polymer 26:1638
38. Abe A, Takeda T, Hiejima T, Furuya H (2001) Macromolecules 34:6450
39. Friend DG, Ely JF, Ingham H (1989) J Phys Chem Ref Data 18:583
40. Sychev VV, Vasserman AA, Zagoruchenko VA, Spiridonov GA, Tsymarny VA (1987) Thermodynamic properties of methane. National standard reference data service of the USSR. A series of property tables. Hemisphere, Washington
41. Friend DG, Ingham H, Ely JF (1991) J Phys Chem Ref Data 20:275
42. Sychev VV, Vasserman AA, Zagoruchenko VA, Spiridonov GA, Tsymarny VA (1987) Thermodynamic properties of ethane. National standard reference data service of the USSR. A series of property tables. Hemisphere, Washington
43. Zoller P, Walsh D (1995) Standard pressure–volume–temperature data for Polymers. Tech, Lancaster
44. Vargaftik NB (1975) Handbook of physical properties of liquids and gases: pure substances and mixtures, 2nd edn. Hemisphere, Washington, DC
45. Abe A, Zhou Z, Furuya H (2005) Polymer (in press)
46. Cheng VM, Daniels WB, Crawford RK (1975) Phys Rev B 11:3972
47. Hust JG, Schramm RE (1976) J Chem Eng Data 21:7–11
48. Price C, Evans KA, Booth C (1975) Polymer 16:196
49. Malcolm GN, Ritchie GLD (1962) J Phys Chem 66:852–854
50. Tonelli A (1970) J Chem Phys 52:4749
51. Costantino MS, Daniels WB (1975) J Chem Phys 62:764
52. Marcelja S (1974) J Chem Phys 60:3599
53. Emsley JW, Luckhurst GR, Stockley CP (1982) Proc R Soc Lond Ser A 381:117
54. Samulski ET, Dong RY (1982) J Chem Phys 77:5090
55. Luckhurst GR (1985) In: Emsley JW (ed) Nuclear magnetic resonance of liquid crystals. Reidel, Dordrecht, p 53
56. Counsell CJR, Emsley JW, Luckhurst GR, Sachdev HS (1988) Mol Phys 63:33
57. Forster P, Fung BM (1988) J Chem Soc Faraday Trans 2 84:1083
58. Abe A, Furuya H (1988) Mol Cryst Liq Cryst 159:99
59. Abe A, Kimura N, Nakamura M (1992) Makromol Chem Theory Simul 1:401
60. Luckhurst GR, Gray GW (eds) (1979) The molecular physics of liquid crystals. Academic, New York
61. Ballauff M (1986) Ber Bunsenges Phys Chem 90:1053
62. Orendi H, Ballauff M (1989) Liq Cryst 6:497
63. Orendi H, Ballauff M (1992) Ber Bunsenges Phys Chem 96:96
64. Oweimreen GA, Lin GC, Martire DE (1979) J Phys Chem 83:2111
65. Oweimreen GA, Martire DE (1980) J Chem Phys 72:2500
66. Kronberg B, Gilson DFR, Patterson D (1976) J Chem Soc Faraday Trans 2:1673–1686
67. Samulski ET (1980) Ferroelectrics 30:1980
68. Gochin M, Zimmermann Z, Pines A (1987) Chem Phys Lett 137:51
69. Janik B, Samulski ET, Toriumi H (1987) J Phys Chem 91:1842
70. Gochin M, Pines A, Rosen ME, Rucker SP, Schmidt C (1990) Mol Phys 69:671
71. Photinos DJ, Samulski ET, Toriumi H (1990) J Phys Chem 94:4688, 4694
72. Photinos DJ, Samulski ET, Terzis AF (1992) J Phys Chem 96:6979
73. Sasanuma Y, Abe A (1991) Polym J 23:117
74. Rosen ME, Rucker SP, Schmidt C, Pines A (1993) J Phys Chem 97:3858
75. Abe A, Iizumi E, Sasanuma Y (1993) Polym J 25:1087
76. Alejandre J, Emsley JW, Tidesley DJ, Carlson P (1994) J Chem Phys 101:7027

77. Sasanuma Y (2000) Polym J 32:883, 890
78. Suzuki A, Miura N, Sasanuma Y (2000) Langmuir 16:6317
79. Sasanuma Y, Nishimura F, Wakabayashi H, Suzuki A (2004) Langmuir 20:665
80. Abe A, Ballauff M (1991) In: Ciferri A (ed) Liquid crystallinity in polymers. VCH, New York
81. Yoon DY, Bruckner S (1985) Macromolecules 18:651
82. Samulski ET (1985) Faraday Discuss Chem Soc 79:7
83. Yoon DY, Bruckner S, Volksen W, Scott JC (1985) Faraday Discuss Chem Soc 79:41
84. Griffin AC, Samulski ET (1985) J Am Chem Soc 107:2975
85. Bruckner S, Scott JC, Yoon DY, Griffin AC (1985) Macromolecules 18:2709
86. Volino F, Ratto JA, Galland D, Esnault P, Dianoux AJ (1990) Mol Cryst Liq Cryst 191:123
87. Sherwood MH, Sigaud G, Yoon DY, Wade CG, Kawasumi M, Percec V (1994) Mol Cryst Liq Cryst 254:455
88. Vorlander D (1927) Z Phys Chem 126:449
89. Kelker H, Halz R (1980) Handbook of liquid crystals. Verlag Chemie, Weinheim
90. Roviello A, Sirigu A (1982) Makromol Chem 183:895
91. Abe A, Furuya H, Nam SY, Okamoto S (1995) Acta Polym 46:437
92. Blumstein RB, Stickles EM, Gauthier MM, Blumstein A, Volino F (1984) Macromolecules 17:177
93. Abe A, Nam SY (1995) Macromolecules 28:90
94. Abe A, Furuya H, Shimizu RN, Nam SY (1995) Macromolecules 28:96
95. Abe A, Takeda T, Hiejima T, Furuya H (1999) Polym J 31:728
96. Emsley JW (1985) In: Emsley JW (ed) Nuclear magnetic resonance of liquid crystals. Reidel, Dordrecht, p 379
97. Ellis DM, Bjorkstam JL (1967) J Chem Phys 46:4460
98. Rowell JC, Phillips WD, Melby LR, Panar M (1965) J Chem Phys 43:3442
99. Abe A (1992) Macromol Symp 53:13
100. Shimizu RN, Kurosu H, Ando I, Abe A, Furuya H, Kuroki S (1997) Polym J 29:598
101. Shimizu RN, Asakura N, Ando I, Abe A, Furuya H (1998) Magn Reson Chem 36:S195
102. Furuya H, Iwanaga H, Nakajima T, Abe A (2003) Macromol Symp 192:239
103. Hiejima T, Seki K, Kobayashi Y, Abe A (2003) J Macromol Sci B42:431
104. Abis L, Arrighi V, Cimecioglu AL, Higgins JS, Weiss RA (1993) Eur Polym J 29:175
105. Abe A, Shimizu RN, Furuya H (1994) In: Teramoto A, Kobayashi M, Norisuye T (eds) Ordering in macromolecular systems. Springer, Berlin Heidelberg New York, p 139
106. Furuya H, Dries T, Fuhrmann K, Abe A, Ballauff M, Fischer EW (1990) Macromolecules 23:4122
107. Furuya H, Abe A, Fuhrmann K, Ballauff M, Fischer EW (1991) Macromolecules 24:2999
108. Furuya H, Okamoto S, Abe A, Petekidis G, Fytas G (1995) J Phys Chem 99:6483
109. Kobayashi N (2004) Master's thesis. Tokyo Polytechnic University
110. Walsh DJ, Dee GT, Wojtkowski PW (1989) Polymer 30:1467
111. Fakhreddine YA, Zoller P (1994) J Polym Sci B Polym Phys 32:2445
112. Orwoll RA, Sullivan VJ, Campbell GC (1987) Mol Cryst Liq Cryst 149:121
113. Chang R (1977) Solid State Commun 14:403
114. Armitage D, Price FP (1977) Phys Rev A 15:2496
115. Stimpfle RM, Orwoll RA, Schott ME (1979) J Phys Chem 83:613
116. Dunmur DA, Miller WH (1979) J Phys 40:C3–141
117. Luckhurst GR (1988) J Chem Soc Faraday Trans 2 84:961
118. Maeda Y, Furuya H, Abe A (1996) Liq Cryst 21:365

119. Abe A, Hiejima T, Takeda T, Nakafuku C (2003) Polymer 44:3117
120. Temperley HNV (1956) J Res Natl Bur Stand 56:67
121. Hardouin F, Sigaud G, Achard MF, Brulet A, Cotton JP, Yoon DY, Percec V, Kawasumi M (1995) Macromolecules 28:5427
122. Imai M, Kaji K, Kanaya T, Sakai Y (1995) Phys Rev B 52:12696
123. Grasruck M, Strobl G (2003) Macromolecules 36:86
124. Meille SV, Allegra G (1995) Macromolecules 28:7764
125. Maier W, Saupe A (1959) Z Naturforsch 14a:882
126. Maier W, Saupe A (1960) Z Naturforsch 15a:287
127. Flory PJ, Ronca G (1979) Mol Cryst Liq Cryst 54:289, 311
128. Frank HS (1945) J Chem Phys 13:495
129. Hildebrand JH, Scott RL (1962) Regular solutions. Prentice-Hall, New Jersey
130. Prigogine I, Trappeniers N, Mothot V (1953) Discuss Faraday Soc 15:93
131. Flory PJ, Orwoll RA, Vrij A (1964) J Am Chem Soc 86:3507, 86:3515
132. Simha R, Somcynsky T (1969) Macromolecules 2:342
133. Nose T (1971) Polym J 2:124
134. Flory PJ (1965) J Am Chem Soc 87:1833
135. Flory PJ (1989) Res Soc Symp Proc 134:3
136. Yoon DY, Flory PJ (1989) Res Soc Symp Proc 134:11
137. Luckhurst GR (1985) In: Chapoy LL (ed) Recent advances in liquid crystal polymerization. Elsevier, London, p 105

Adv Polym Sci (2005) 181: 153–177
DOI 10.1007/b107180

Published online: 30 June 2005

Motional Phase Disorder of Polymer Chains as Crystallized to Hexagonal Lattices

P. Sozzani (✉) · S. Bracco · A. Comotti · R. Simonutti

Department of Materials Science, University of Milano – Bicocca, Via R. Cozzi 53, 20125 Milano, Italy
Piero.sozzani@unimib.it

Abstract Chain-dynamics of flexible synthetic macromolecules in the crystalline state has been clarified by multinuclear solid-state NMR observations. Massive motional phenomena are shown when chains are crystallized in the hexagonal phase. Several examples have been surveyed, where motion is induced by high temperature and pressure or favored by inclusion of moderate amounts of comonomer units in the main chain and by the use of monomers containing mobile side chains. Also, motional phenomena of fast spinning and large librations of the macromolecules are promoted when the isolated chains are confined to nanochannels formed in hexagonal crystal structures of suitable matrices. The motional behavior can be modulated by the topology and the interactions in the unusual environments. This knowledge provides information about polymers in close contact with heterogeneous interfaces, such as polymers in nanocomposites or at surfaces.

Keywords Polymer crystallization · NMR of polymers · Polyethylene · Hexagonal phases · Nanostructured materials · Confined polymers · Crystal engineering · Nanochannels

1 Introduction

NMR is a method of choice for understanding the details of the conformations and dynamics of semi-crystalline macromolecules in their crystalline and mesomorphic state and interphases [1–7]. It complements other structural methods for recognizing chain arrangements and aggregation states, and is particularly suitable for understanding intricate morphologies. Moreover, it is largely known that the dynamics of the process of phase aggregation or chain rearrangements in the time regime of nanoseconds to seconds are compatible with the wide frequency spectrum that NMR explores. NMR provides direct evidence of motional mechanisms and disordered states that are difficult to be recognized unambiguously by XRD. However, in-depth works devoted to determining polymer structure in the crystalline phase by NMR are still relatively few compared to those performed by classical structural methods. A practical reason is the limited number of NMR laboratories dedicated to the study of supramolecular and macromolecular structures.

In the present overview we show that NMR contributed substantially to the comprehension of the mesomorphism of hexagonal phases and the motional behavior of polymer chains in crystals.

2
Extended Chain Conformations, Hexagonal Phases and Rotational Disorder of Flexible Polymers

The concepts of extended chain conformations, hexagonal phases and rotational disorder are frequently linked one to another. In fact, flexible polymer chains, by spinning motion and libration about their chain axes, may generate ideal cylinders that find hexagonal packing as the easiest way to assemble. The hexagonal phase, due to its intrinsic mobility, is favorable for obtaining lamella thickening up to the limit of the thermodynamically stable extended-chain morphology. On the other hand, when drawn in the extended-chain morphology, the chains lose specific orientations in planes perpendicular to the chain axis and increase the hexagonal phase fraction [8–10]. The hexagonal phase is also involved in the onset of the crystallization process [11, 12]. Thus, one of the key-points of this family of phenomena is likely the reorientational diffusion motion about the polymer axis. In fact, under suitable conditions, the existence of anisotropic motions in the solid can be preferred over the larger freedom of the melt state, as happens, in particular, when pressure counterbalances thermal agitation at elevated temperatures.

The transition of polyethylene from the orthorhombic to the hexagonal phase was observed at very high pressures (3000–5000 bar) by proton NMR and nuclear magnetic relaxation time measurements: the phase diagram derived from this spectroscopy is in substantial agreement with DSC data. According to these results (see Fig. 1), the hexagonal phase can be observed above the triple point at 490 K and 3000 bar [13, 14].

Solid-echo deuterium NMR spectra provided unique evidence about motions occurring in the crystals at those extreme conditions. Among other methods reported in the literature, only Raman spectroscopy could identify the conformationally disordered chains in the crystals at high pressure and temperature. Deuterium line-shape analysis offers the possibility to calculate spectral profiles as derived by reorientation models of the C – D bonds. Diffusive rotation of polyethylene chains as rigid bodies implies that C – D vectors explore randomly any orientation on planes perpendicular to the chain axis. Thus, pure rotation of the polyethylene chains about their axes in the all-*trans* conformation should produce a Pake powder pattern reduced to half the width of the static spectrum, showing a splitting between singularities of 62 KHz. The actual lineshape presents a splitting between singularities sensibly smaller (about 50 KHz) (Fig. 2) and indicates an additional reorientation mechanism obtained from the wobbling of C – D bonds 23° out of plane. The reorientation must be fast compared to the 10^6 Hz frequencies of the deuterium NMR experiment.

Thus, chain twists and distortions from the *trans* planar conformations are to be included in the model for understanding the spectral output of the hexagonal phase. Chain reorientation can also be present in phases with a degree of order higher than that of the hexagonal phase. For instance, a mechanism of rapid 180° chain flips about the chain axis has been described by NMR and makes compatible an orthorhombic phase with the presence of chain motion [15].

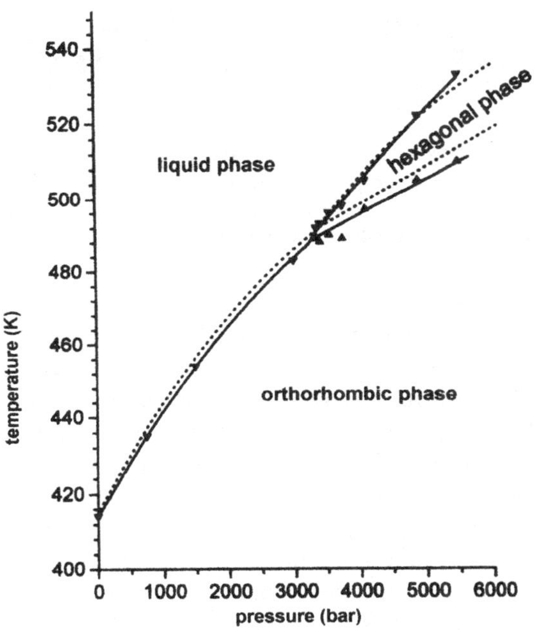

Fig. 1 Polyethylene phase-diagram: ^{1}H solid state NMR data (*solid lines*) according to [13] compared with DSC data (*dotted lines*)

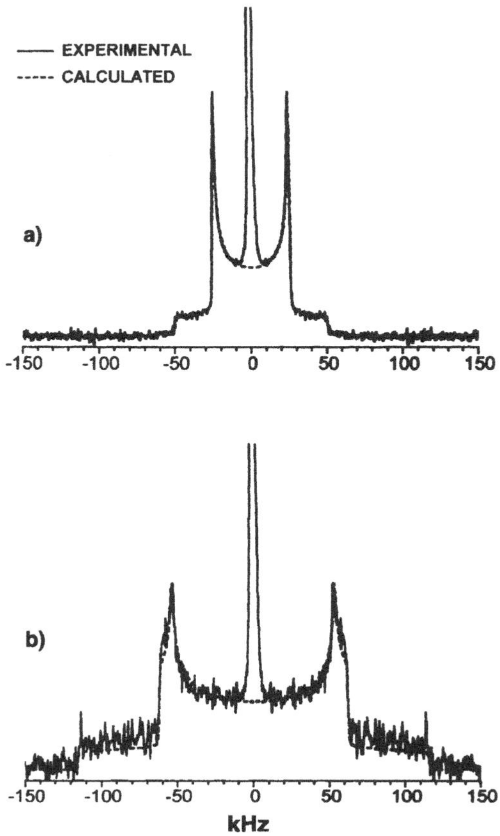

Fig. 2 Static deuterium NMR spectra at 4900 bar of polyethylene (**a**) in the hexagonal phase and (**b**) in the orthorhombic phase, showing the mobility of the hexagonal phase, as indicated by the shrinkage of the singularities to about 50 kHz. The isotropic signal at the center of the spectrum represents the mobile amorphous phase

Linear alkanes are known to produce similar diffusional motions in the hexagonal phase under much milder conditions and static-to-rotator phase transitions occur at lower temperature and pressure. *Trans/gauche* conversions have been observed in the short *n*-alkane chains for the larger freedom induced by the chain ends. The conformational interconversion and the spinning of the chains have been shown in many instances by MAS NMR [16–20].

Ethylene sequences in copolymers of ethylene with α-olefins can crystallize and show an interesting polymorphic behavior, similar to linear alkanes. Monoclinic phases are commonly induced when comonomers bear bulky side-groups and their mole fraction is more than 10% [21], but the copolymers form hexagonal mobile phases if pendent groups are moderately bulky, like methyl groups. Hexagonal aggregates can exist only at low temperature if the average length of ethylene homosequences are short: they were observed by NMR in ethylene-*ran*-propylene copolymers with an ethylene monomer-unit mole content ranging from 50%, as the lower limit, to about 75% [22].

By lowering the temperature below room temperature, *trans/gauche* balance changes in the ethylene sequences and aggregation to an ordered *trans*-rich phase occurs. The crystallization to this phase depends on the ethylene content and happens at temperatures ranging from room temperature to – 50 °C (just above the glass transition temperature) (Fig. 3).

At 75% ethylene content melting/crystallization occurs at 9 °C (Fig. 3). Small-size aggregates are formed in the crystallization process. Their existence is quite intriguing since they represent short living species (lifetime of 3 ms) that precede crystallite growth in the hexagonal modification. Propylene monomer units are clearly included in the aggregates, since triads containing ethylene and one propylene unit (e.g. propylene-ethylene-ethylene PEE triad) show the transition of *gauche*-rich to *trans*-rich chains at about the same temperature as homopolymeric ethylene triads (ethylene-ethylene-ethylene EEE triad) (Fig. 4).

The lack of hysteresis shows the intimacy to the amorphous phase (Fig. 5). Both chain mobility and fast carbon spin-lattice relaxation times ($T_1(^{13}C) \approx 1$ s) are promoted by the random distribution of the propylene units that push the neighboring chains apart. The copolymer chain segment behavior in the aggregates is dynamic, similar to that of polyethylene as arranged in the hexagonal packing and to that of polyethylene mesophases confined to crystalline nanochannels as described below.

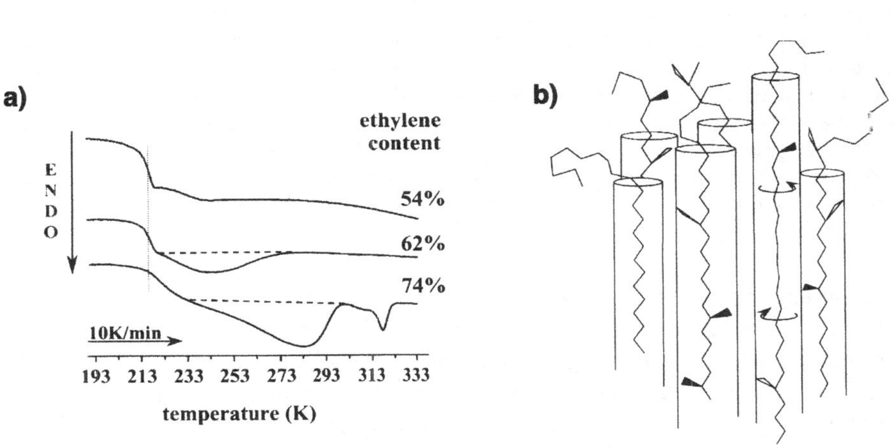

Fig. 3 (**a**) DSC traces of ethylene-propylene (EP) copolymers with different ethylene fraction: the melting of the crystalline aggregates is shown by a wide endotherm transition increasing with ethylene content (**b**) Sketch of the aggregates in the hexagonal form that include isolated monomer units of propylene within ethylene homosequences. Large librations can reorient the copolymer chains within the cylinders, inducing low-density packing

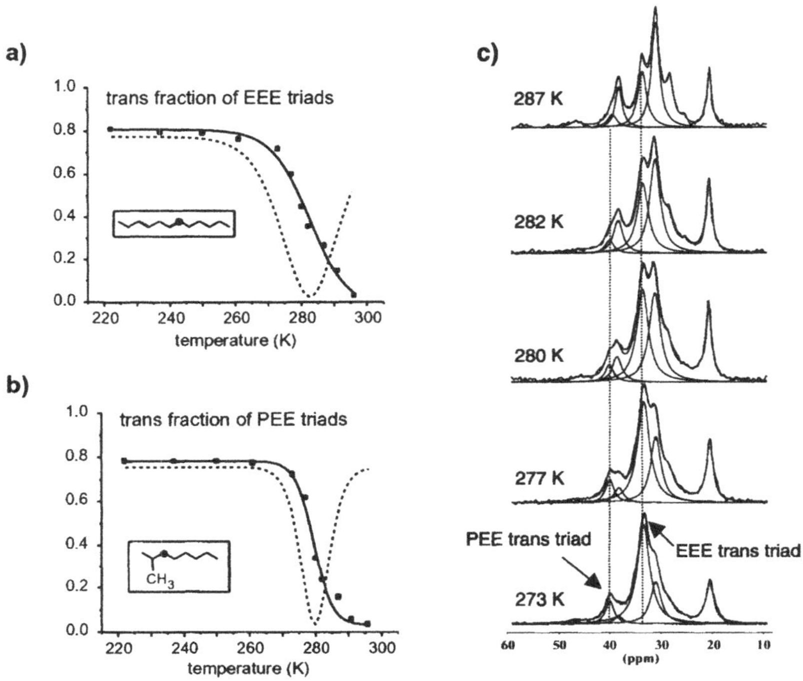

Fig. 4 (**a**) *trans* fraction of the EEE homosequences of ethylene-*ran*-propylene copolymer with 75% ethylene content as a function of temperature. The *trans* fraction of EEE triads was determined from the analysis of the peak areas at 33.6 and 30.8 ppm (**b**) *trans* fraction of the PEE heterosequences of ethylene-*ran*-propylene copolymer with 75% ethylene content as a function of temperature. The heterosequences contain a methyl as a substituent in the α position with respect to the observed methylene, and the *trans* fraction of the triads was determined from the deconvolution of the CP MAS NMR spectra performed at varying temperatures. The first derivatives of the curves are also reported (*dotted lines*) to localize precisely the transition (**c**) ^{13}C CP MAS NMR spectra of the ethylene-*ran*-propylene copolymer with 75% ethylene content recorded with a contact time of 2 ms at variable temperatures

The aggregates are hardly detectable by those techniques that are sensitive to long range order. Density fluctuations in the amorphous phase might be reinterpreted in some instances as labile crystalline aggregates living a few milliseconds. At 75% ethylene content, a small fraction of aggregates can grow to form crystallites stable above room temperature and melting at 45 °C [22]; at 80% ethylene content the sample exhibits at room temperature 14% of crystalline hexagonal phase; a further increase of the average length of ethylene homosequences leads to crystallization to the conventional orthorhombic crystals, as shown by the NMR profile in static carbon spectra [23].

Another simple case of a flexible chain without side groups, that is able to crystallize in the hexagonal form, is 1,4-*trans*-polybutadiene. *Cis/skew*

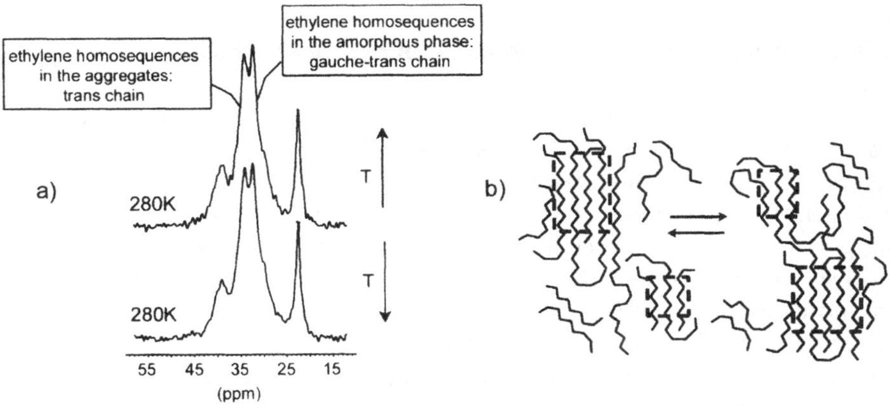

Fig. 5 (**a**) ^{13}C CP MAS NMR spectra recorded at 280 K of ethylene-*ran*-propylene copolymer with 75% ethylene content, showing at 280 K an equivalent amount of ethylene homosequences in the crystalline aggregates and in the amorphous phase. The spectra are identical on heating and cooling, indicating the lack of hysteresis in the crystallization process (**b**) A simple model of the ethylene homosequences: fast exchange between the aggregates and the amorphous phase is reported

conformations, close to the double bonds and easy to be converted one to another, play the role of articulated joints permitting anisotropic motions sustained by thermal agitation. Unlike polyethylene, the transition to a mobile "rotator" phase occurs under the mild condition of atmospheric pressure and above 70 °C (Fig. 6).

Above 70 °C carbon chemical shifts, relaxation times and deuterium NMR are in agreement with the model of a highly mobile chain spinning in a cylinder, to be compared with the polymer chains in inclusion compounds, as discussed later [24, 25].

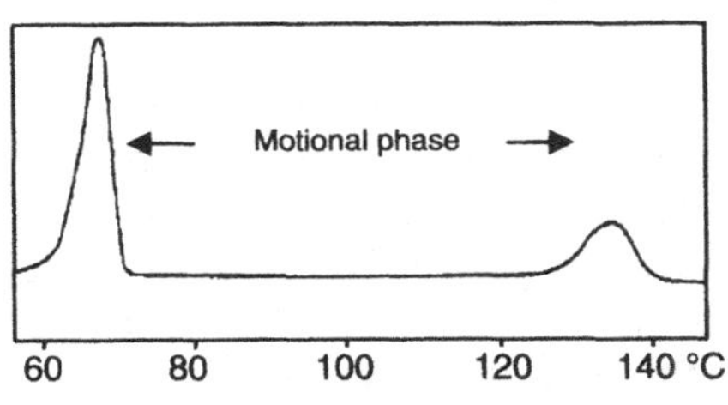

Fig. 6 DSC analysis of 1,4-*trans*-polybutadiene in the bulk phase. The two endotherms correspond to the crystal–crystal transition at 70 °C and to the melting point at 133 °C

3
Side-Chain Flexibility and the Formation of Hexagonal Phases

Flexible side chains are at the origin of the generation of columnar mesophases and hexagonal phases. Crowded and bulky substituents on the main polymer chain may generate the shape of a cylindrical rod and induce the aggregation to hexagonal phases. If specific interactions between neighboring polymer rods are scarce, two kinds of motion can occur: the first kind involves side groups and the second kind a spinning rotation of the chain around its axis. The entire polymer chain can undergo rapid rotational diffusion or large librations about the *c* axis. Solid–solid phase transitions are associated with the occurrence of motion of the side chains. The population of substituents grafted onto the main chain can become crowded, reducing the main chain flexibility. Polyphosphazenes that bear two substituents linked to phosphorous by $P-O$ or $P-C$ bonds, are excellent as case studies for this group of phenomena and NMR spectroscopy can give precious details about the conformations in the static and dynamic modifications. The first high resolution solid state NMR of polyphosphazenes was performed by Haw et al. [26] (Fig. 7), although wide-line studies had been previously reported [27]. These polymers show onset of motions at the solid–solid transition temperatures (T_c). Major changes of chemical shifts, relaxation times and line-width occur at the transition temperature in the class of polyalkoxy- and polyaryloxyphosphazenes.

This implies a fast, and complete, reorientation of the side groups, as in the molten state. Nevertheless, local anisotropy of motion is retained up to the much higher isotropization temperature (T_i). Great heat absorption is involved in the T_c transition, while a minor DSC signal is detected at the melting. Such a dramatic transformation of a large mass of the solid at T_c temperature supports the definition of a crystalline-to-mesophase thermotropic transition.

A second family of polymers which are of interest because of their crystalline state behavior bear alkyl groups directly attached to the phosphazene backbone by a carbon–phosphorous bond. In fact, at ambient temperature a multiplicity of side-group conformations could be detected, and a solid–solid transition is produced at the onset of fast motions about the $P-CH_2$ bonds. In these polymers the T_c transition occurs at moderate temperature, together with the absorption of a small amount of heat. Longer alkyl chains attached to the backbone can lead to changes in the carbon–carbon and $P-CH_2$ bond conformations on the side chains [28]. XRD recognizes two crystalline forms of polydiethylphosphazene: form I, obtained as polymerized after precipitation from poor solvents, and form II, obtained after melting and crystallization. Form I presents a solid–solid transition at 318 K. The ^{13}C MAS NMR spectra produced by the ethyl groups are informative about the

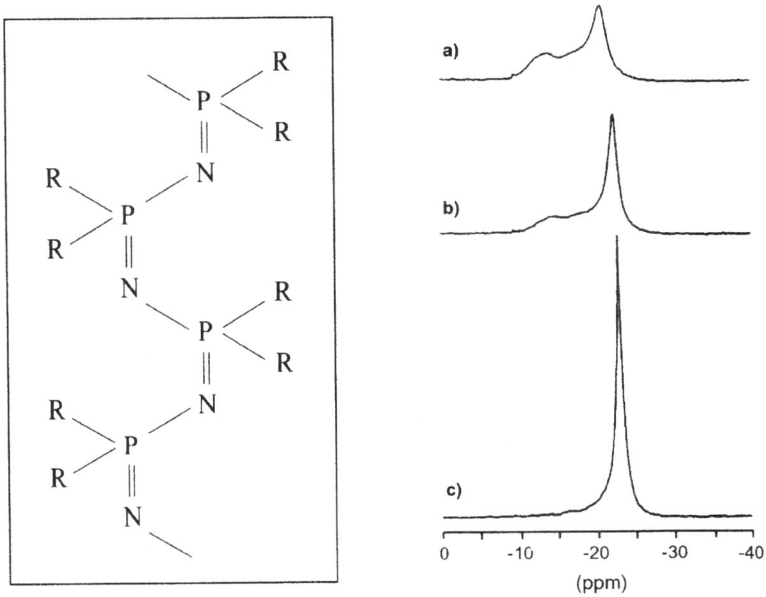

Fig. 7 Schematic representation of polyphosphazene chains; "R" indicates: alkyl, alkoxy or aryloxy substituents. ^{31}P MAS-NMR of poly[bis(4-ethylphenoxy)phosphazene] at (**a**) 23 °C (**b**) 80 °C and (**c**) 120 °C

origin of the transition. Modification I of polydiethylphosphazene shows at 293 K a multiplicity of peaks in the ^{13}C spectrum of the methyl and methylene region (Fig. 8a).

The ethyl groups are arranged at room temperature in four stable conformations identified by four signals in the methyl region. An average chemical shift is measured above the transition temperature T_c, at 323 K, due to the rapid conformation interchange. However, the signals were not assigned until a comparison had been made with the corresponding cyclic trimer (hexaethylcyclotriphosphazene) (Fig. 8b). The cyclic trimer is a crystalline material that can be considered a model compound of the open chain polymer. Like the polymer, the cyclic trimer shows a phase transition at 313 K and four lines in the methyl region of its ^{13}C MAS NMR spectrum at 293 K (Fig. 8b).

The intensity distribution of the lines showing the internal ratio of 2 : 1 : 1 : 2 is consistent with a molecule bearing 6 ethyls, there being double intensity of two signals due to the identical conformations related by a plane of symmetry in the hexagonal unit cell. The chemical shifts of the methyl lines were assigned to four conformations according to the γ-gauche effect [28]: essentially, two γ-gauche interactions with nitrogens explain downfield signals, while one γ-gauche interaction with nitrogen and one with carbon explain the upfield signals (Fig. 9a). The temperature increase brings a narrowing in the range of methyl spectra, and a signal at the weighted average of the chem-

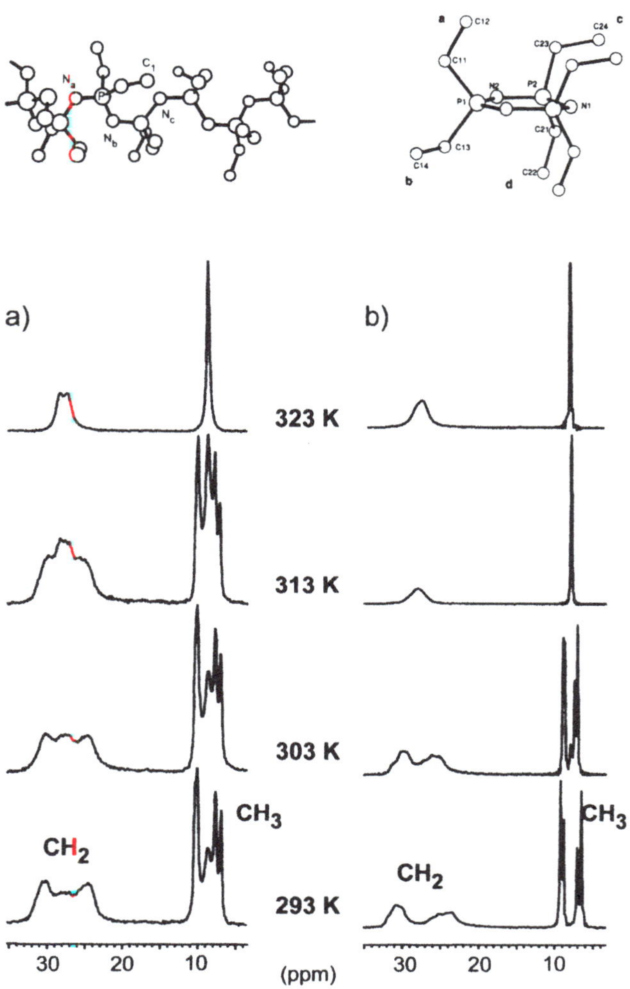

Fig. 8 ^{13}C CP MAS NMR spectra of: (**a**) polydiethylphosphazene and (**b**) hexaethylcyclotriphosphazene: the cyclic model of the polymer. The conformational arrangements of the ethyl groups are subjected to a fast exchange in the NMR timescale above the solid–solid transition

ical shifts appears and grows. Above the transition only the average signal survives at the center of the spectrum.

At room temperature four lines are recognized also in the methylene spectrum, each split by coupling to a ^{31}P atom (Fig. 9b). In fact, the methylene spectrum retains the information about the *J*-coupling to the ^{31}P, which is a 100% natural abundance spin-active nucleus. Thus, the ^{31}P–^{13}C *J*-couplings as a function of conformations were established.

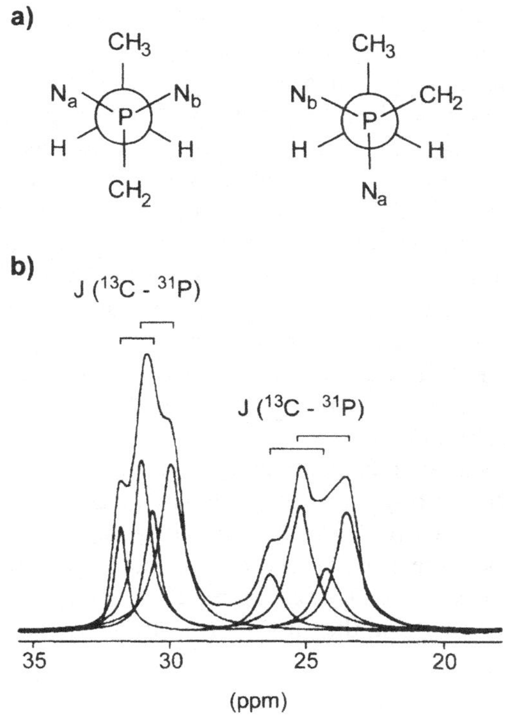

Fig. 9 (**a**) Newman projections, along the P – C bond of two ethyl group conformations in polydiethylphosphazenes (**b**) ^{13}C MAS NMR spectrum of the methylene region of the hexaethylcyclotriphosphazene at room temperature, indicating the dependence of ^{13}C–^{31}P *J*-couplings on the conformations

The transition T_c can not be detected by XRD as it involves the change from static to dynamic disorder of the ethyl groups. The onset of motion in the ethyl groups provides a good explanation for the high melting point (isotropization temperature of 503 K), this being due to the modest amount of entropy gained for the transformation from the mobile solid to the isotropic liquid. However, these experiments did not demonstrate whether or not the side groups could undergo complete reorientation about the P – CH_2 bond. This reorientation should imply cooperation of neighboring groups (Fig. 10a). The motional mechanism was highlighted by a rotor-synchronized triple resonance NMR experiment called TRAPDOR suitable for the evaluation of the shortest distances between carbons and nitrogens in the low- and high-temperature modifications [29]. The experiment is sensitive to the inverse of the third power of the distance. The agreement between the measured distances to those calculated by the molecular modeling demonstrate that, in the high-temperature modification the ethyl groups undergo a complete rotation about the $P-CH_2$ bond by exploring even those conformations which produce short carbon-to-nitrogen distances (Fig. 10b).

Deuterium NMR of polyphosphazenes substituted by perdeutero or selectively deuterated ethyls was interpreted by threefold jumps among the three conformational minima [30]. Molecular mechanics models gave support to a concerted motion of the ethyls surrounding the main chain, which is thus embedded in a "soft" cylindrical environment, fitting a hexagonal lattice.

Polydipropylphosphazene and the cyclic trimer hexapropylcyclotriphosphazene can possibly have a number of conformations for the flexible side chains. This poses a challenge to NMR for eliminating the phosphorous-to-carbon couplings: the ^{13}C MAS spectra were recorded by a double high-power decoupling from ^{31}P and ^{1}H. This yielded high resolution spectra: that of hexapropylcyclotriphosphazene is reported together with a picture of the molecule and the assignments (Fig. 11).

The conformational arrangements and the assignments were established by comparison with the resolution of single crystal X-ray diffraction analysis [31]. The splitting of each carbon atom into three signals of equal intensity, as clearly seen in the methyl region and for the methylenes attached to phosphorous atoms, derives from a C_2 axis intersecting the molecule and passing through the nitrogen N(2) and the phosphorous P(1) atoms. The independent

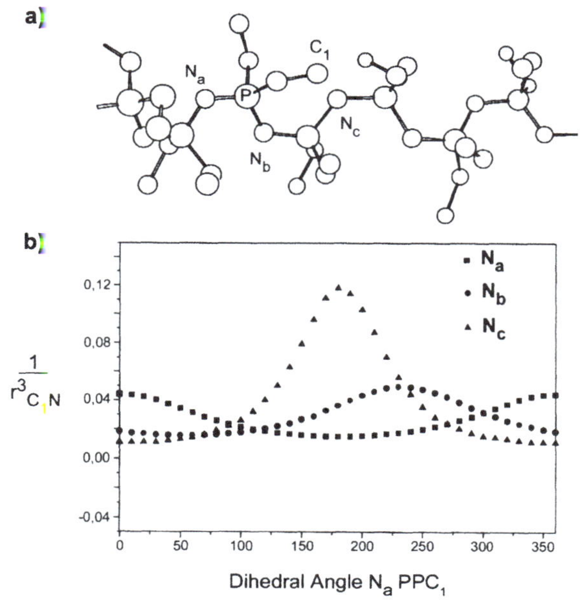

Fig. 10 (**a**) Polydiethylphosphazene chain displaying the ethyl group conformations (**b**) Plot of the inverse third power of the distance between methyl C_1 and nitrogens N_a, N_b, N_c, as a function of the dihedral angle $N_a-P-C-C_1$

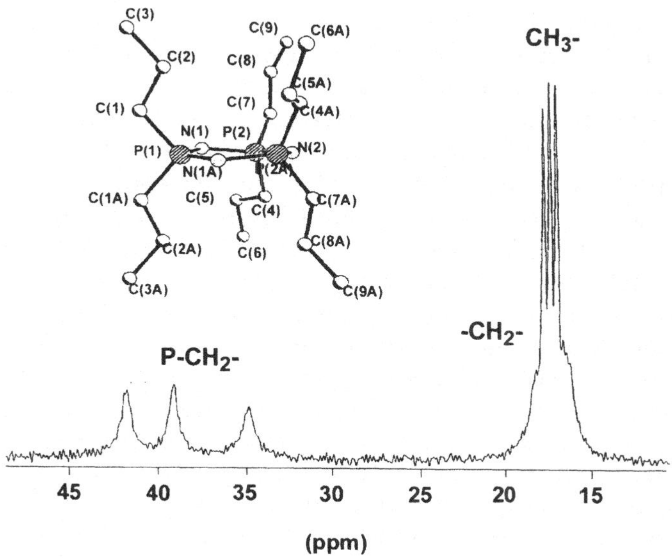

Fig. 11 ^{13}C MAS NMR spectrum of hexapropylcyclotriphosphazene at room temperature with simultaneous ^{1}H and ^{31}P decoupling: the chemical shift of three conformations for each carbon atom is detected

carbon atoms of each species are in fact reduced from the total number of 6 to the number of 3. The carbon–carbon and carbon–nitrogen γ-gauche effects explain the three $CH_2 - P$ signals [28]. Again it is the full identification of the stable conformations that highlights the thermotropic transition of the cyclic trimer crystal and of the polymer.

4 Assembly of Linear Polymers with Hosts, Fabricating Hexagonal Lattices

Some low molecular mass molecules of suitable shape and symmetry can assemble spontaneously with linear polymers to form inclusion compounds. The crystal architectures where polymers sit are sustained by a network of weak forces. These adducts generally show a hexagonal lattice that melt at temperatures higher than those of the bulk polymers [32]. The high melting temperatures are explained by the relevant enthalpy of association and by high mobility of the polymer chains in the solid state, such as large rotations or librations. Flexible polymers entrapped in nanochannels explore limited diffusional modes where translational diffusion and chain bending are prevented. The crystalline host partially transfers to the chains its anisotropic arrangement and, to some degree, its order.

Polymers organized in this unusual state behave as hexagonal mesophases similar to those observed in the bulk at suitable temperature/pressure conditions and adopt the extended chain conformation. Polymers included in nanochannels were discovered a few decades ago [33, 34], but the mesomorphic properties and the stabilizing interactions were established much later by advanced spectroscopic techniques [35–41]. The preparation of novel macromolecular adducts, melting at temperatures as high as 350 °C and sustained by CH$\cdots\pi$ intermolecular interactions, has been a success of supramolecular chemistry in fabricating high performance nanostructured materials [42].

The techniques for including the polymers are the crystallization from the melt with suitable hosts, and the crystallization from a solvent common to the polymer and the host by self-assembly. In some cases simple milling of the host and the guest leads to the inclusion. A particular case is that of large molecular rings, that can accept linear alkanes and polymer chains like a thread in the eye of a needle, forming *rotaxanes* and *pseudorotaxanes*: the better known examples are formed by cyclodextrins [43–45].

An alternative route to the formation of polymer nanocomposites is the polymerization of the monomers organized in solid media to high molecular mass polymers. Linear macromolecular chains in nanochannels, either polymerized in situ or confined by self-assembly, exhibit the special properties of isolated and elongated chains. The conformations become similar to those of bulk-phase polymorphs existing at special pressure and temperature conditions or aggregated following peculiar thermal histories. Thus, this paragraph is intended to describe the behavior of the polymer guests in the hexagonal structure and the host–guest properties are compared with the bulk polymer properties. A typical DSC trace showing the formation of a macromolecular inclusion compound (IC) by self-assembly is reported in Fig. 12 for PE with the crystalline matrix of tris(-o-phenylene)dioxycyclotriphosphazene (TPP). The special properties of TPP are shown in detail later.

The crystallization of the adduct is evidenced by an exotherm with an enthalpy of formation of 140 J mol^{-1} followed by the congruent melting at a temperature higher than the melting temperature of both components. In a second example the thermal degradation of the polymer (PEO) follows immediately the melting of the inclusion compound, suggesting that the polymer is well protected from degradation as long as the capsule of the host structure survives [46]. Thus, polymer chains can be observed in the solid state at temperatures much higher than the melting temperature and even above the degradation temperature of the bulk polymer.

Polymers show special thermal properties not only in the nanostructured materials but even upon removal of the host. The extended-chain conformation is retained after extraction by solvents and has an impact on the bulk physical properties. In fact, bad solvents of the polymer favor the collapse of the polymer chains next to and parallel to one another, thus inducing the high-melting morphology typical of extended-chain polymers [47].

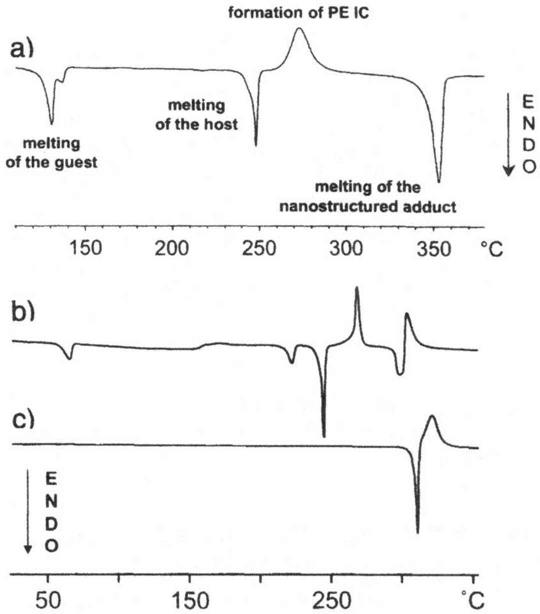

Fig. 12 (**a**) Differential scanning calorimetry DSC trace observed in the first run of TPP/PE mixture; a peak at 350 °C is detected indicating the melting of the nanostructured adduct (**b**) DSC trace for a mixture of TPP and PEO (first run) showing the melting of both the pure polymer (65 °C) and the matrix (250 °C) followed by the congruent melting of the inclusion compound at 300 °C. The exothermic peak at 270 °C indicate the formation of the inclusion compound (**c**) DSC trace of the TPP/PEO IC obtained by co-crystallization showing the congruent melting of the inclusion compound and the degradation process

Motional and mesomorphic states were found in 1,4-*trans*-polybutadiene included in nanochannels after formed in situ by solid state polymerization in the matrix of perhydrotriphenylene (PHTP) (Fig. 13) [48].

^{13}C MAS NMR and solid echo ^{2}H NMR together with the relaxation time measurements [25, 36], suggested great similarities to the polymer in the bulk of the crystalline form II, existing above 340 K (see Fig. 6). The motional behavior is not quenched even at 115 K. The motional frequency component in the observed frequency regime of 10^8 Hz provide short carbon spin-lattice relaxation times (T_1s of 10 s): not so for the matrix that still shows relaxation times in the hundreds of seconds, even if reduced by a factor of 2 compared to the pure matrix. This is an indication for a modest coupling of the guest motion to host vibrations, confirming the function of robust nanovessels performed by the host crystals. The insertion of substituents on the polymer main chain modifies the host–guest steric interactions, that stop the polymer motion. The consequence is the formation of a static polymer chain

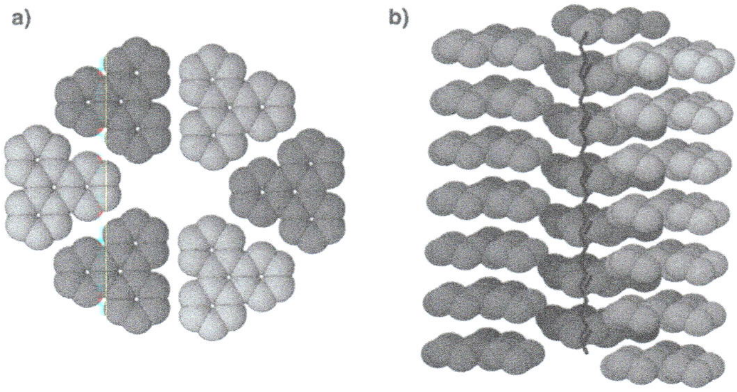

Fig. 13 (**a**) Projection of PHTP crystal structure along the channel axis, the polymer chains are omitted for clarity (**b**) Crystal structure of 1,4-*trans*-polybutadiene in the PHTP inclusion compound

in a nanocomposite structure: examples being 1,4-*trans*-polyisoprene (balata rubber) [38] and 1,4-*trans*-poly(1,3-pentadiene) [49].

A few macromolecular chains were tried for a proper fitting in varied size ideal cylinders simulating nanochannels or nanotubes. The conformations adopted by some polymers in a cylinder were modeled using Rotational Isomeric States treatment: the selection rules are severe for a 5 Å channel [50, 51].

The analogy between squeezing flexible macromolecules in nanochannels and stretching them is apparent. This is a stimulating example of using crystals as nanotechnology tools able to direct forces in specific ways for controlling, among other features, anisotropic phenomena. Thus, nanocrystalline channels can isolate single polymer chains and can impose the hexagonal symmetry. The properties of the isolated chains become similar to the hexagonal phases in the bulk, showing how the channels force the chains to a novel structure. A few novel 'nanophases' could be identified and modulated by the host environment. Urea was the first host applied: the cross section fits closely polymethylene chains and polyoxymethylene chains. XRD shows the existence of matrix-induced phase-transitions. Above the phase transition the polymer apparently experiences a slightly looser constriction, enough to gain motional freedom.

Direct observations of polyethylene embedded in the fully aromatic channels of TPP nanostructured materials was provided by NMR spectroscopy in one- and two-dimensions (Fig. 14) [42].

The unique aromatic environment affects included polymer chains, where specific arrangement of phenylenedioxy rings constitutes the lining of the walls [52]. The methylene units are directed towards the aromatic rings in the intimate nanocomposite material. Thus, a large upfield shift of the resonance frequency is obtained, that derives from the magnetic susceptibility contri-

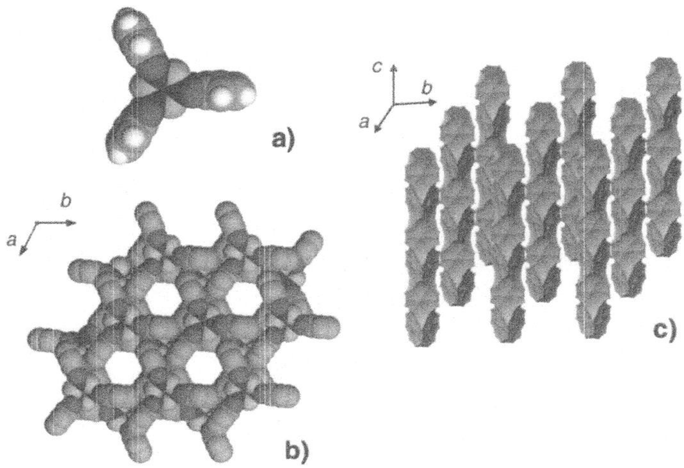

Fig. 14 (**a**) Representation of the tris-o-phenylenedioxycyclotriphosphazene (TPP) molecule with D_{3h} symmetry (**b**) Crystal structure of TPP in the nanoporous hexagonal modification viewed along the channel axis: a double layer of molecules is presented (**c**) Portion of the space described by the center of a sphere of 2.5 Å diameter exploring the empty nanochannels ($3 \times 3 \times 3$ array of unit cells) [52]

butions generated by the aromatic ring of the host facing the polymer chains and proves the intimacy of the components in the nanostructured material.

By the application of Lee–Goldburg decoupling both the hydrogen and carbon spectral dimensions are well resolved [53, 54]. Proper pulse sequences can give 2D spectra containing targeted information without selective deuteration of the single phases. In the 2D heterocorrelated experiment, proton magnetization evolves during t_1 and then is transferred by cross-polarization to the nearby carbons. The correlation signal intensity is a measurement of the distance traveled by the magnetization. The most effective transfer highlights the shortest hydrogen-to-carbon distances. A 2D ^{1}H–^{13}C heterocorrelated (HETCOR) spectrum of PE/TPP hexagonal adduct is shown in Fig. 15.

In the same NMR spectrum the pure polymer resonates at a separate chemical shift value. Host–guest cross peaks are diagnostic of the nanoscale topological relationship, giving an insight into the incommensurate crystal structure. This kind of spectroscopy was extended from polymethylene chains to a number of polymer nanocomposites, including rubbery polymers. The most interesting examples, those are formed with elastomers, where the crystalline adducts act as reinforcement for the elastomeric material [55].

The exceptional resolution in the hydrogen domain and the large magnetic susceptibility was exploited to understand the precise topology of the hydrogen atoms in the crystal structure. In a control NMR experiment with contact times of 1 ms Fig. 1a, only intramolecular correlations of protons to

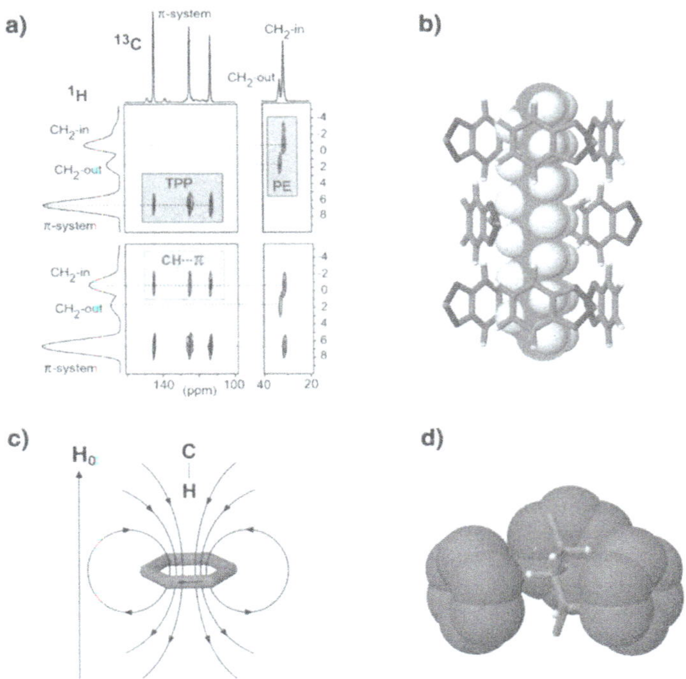

Fig. 15 (**a**) Phase Modulated Lee–Goldburg (^1H–^{13}C) HETCOR NMR spectra of TPP/PE adduct: contact times of 1 ms (above) and 5 ms (below). CH_2-in indicates the PE chains included in TPP nanochannels; CH_2-out indicates excess PE. The CH$\cdots\pi$ interaction between the $-(CH_2)_n-$ of PE and the π-system of TPP is highlighted in the spectrum below (**b**) Extended conformation of $-(CH_2)_n-$ chain wrapped by the TPP paddles that form a 0.5 nm aromatic nanochannel (**c**) Topology of guest hydrogens located at 2.5–2.7 Å above the plane of the benzene rings: the ring currents generate an upfield shift of 2.2 ppm (**d**) A linear aliphatic chain encased in the nanochannel and surrounded at a close contact by the aromatic paddles of TPP [42]

vicinal carbons are active. On the contrary, the 2D spectrum of the adduct width contact times of 5 ms shows a multiplicity of host–guest interactions and a strong upfield shift of the polymer signals in the carbon and hydrogen dimensions [42].

The matrices examined so far produce hexagonal lattices of different cross section and modulate the space available for the included macromolecules. The mechanism of motion of the guest polymer and the motional frequency were carefully measured in the hexagonal lattices. Motions are faster than the Megahertz regime and fall close to the maximum of relaxation rate in PHTP [39], reaching the extreme narrowing limit typical of liquids in TPP [41]. Chain motion is not hindered in TPP because the density of aromatic rings lining the nanochannel is high (three per cross-section) and permits the guest protons to pass smoothly from the influence of one ring to the next. Thus, intriguingly, rotations and librations are allowed, since a spinning polymer chain never falls into a deep energy minimum [42]. Solid-echo deuterium NMR confirms the occurrence of diffusional reorientation about the polymer axis (50 kHz between the singularities), modeled as a complete averaging for fast ($\tau_c < 10^{-6}$ s) on-plane rotations or librations as shown in Fig. 16. T_1 of 12 s of the polymer carbons in TPP, in the extreme narrowing limit typical of liquids, suggests an even faster spinning of the large segments of the polymethylene chains ($\tau_c < 10^{-10}$ s). The exceptional mobility of the polymer chains in the fully aromatic environment creates a unique example of a macromolecular rotor stabilized by soft interactions.

No conformational transitions occur in the inclusion compounds of PE. Strictly speaking, a fast reorientation of high molecular weight flexible polymer as a rigid body is a kind of a contradiction. The flexibility of the chain and the statistical departure of the single carbon–carbon bond conformation from the exact *trans* arrangement must be taken into account. In fact, 15–20° wide librations are tolerated without changing much the internal energy of the molecule. Molecular dynamics simulations performed over nanoseconds depicted some of the motion likely occurring in the tubular space [56]. They confirm large librations and oscillations as the main mechanism of chain reorientation in the hexagonal lattice.

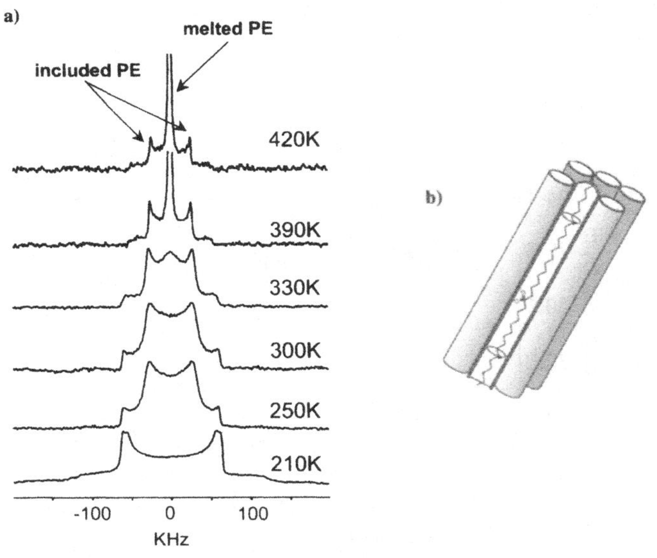

Fig. 16 (**a**) ^{2}H solid-echo NMR spectra of TPP/PE-d_4 adduct collected at variable temperature (**b**) Schematic picture of the PE chain rotating about the polymer axis

In general, no large energy barriers are to be crossed for the motional phenomenon. The activation energy was evaluated by the relaxation values to be as small as 4.5 kJ mol^{-1} in PHTP, lower than a *gauche/trans* interconversion barrier [39]. Large librations of methylenes account for the short carbon spin-lattice relaxation times at room temperature.

Short polymethylene chains with less than 40 carbon atoms when included in PHTP channels show a NMR shift due to kinks of conformations propagating from the chain ends. In TPP the effect is restricted to shorter (< 20 carbon) chains. This is a clear indication for a topological control exerted by each matrix according to the cross-section (Fig. 17).

Thus, cylindrical nanochannels and linear polymers are complementary shapes, in such a way that polymers experience diffusional spinning motion in the channels, as prototypes of molecular rotors [57]. The data suggest that a small channel can even facilitate rotation, by eliminating the occurrence of space-demanding defects. Defects like chain-twists, not involving lateral swelling are rapidly traveling along the chain and are therefore called *twistons*. A facilitated rotational diffusion in straight tubes can be proposed in analogy to the reptation phenomenon described in the amorphous phase [58]. At high temperature the phenomenon intensifies but, not so much as to cause the system to melt. The series of spectra at varied temperatures show a slight narrowing of the singularities in the deuterium spectra at high temperature, well accounted for by a secondary reorientation along the *c*-axis, that is a necessary condition when a torsion of two neighboring bonds occurs. The property of a mesophase is confirmed and brings light to the hexagonal arrangements in the bulk.

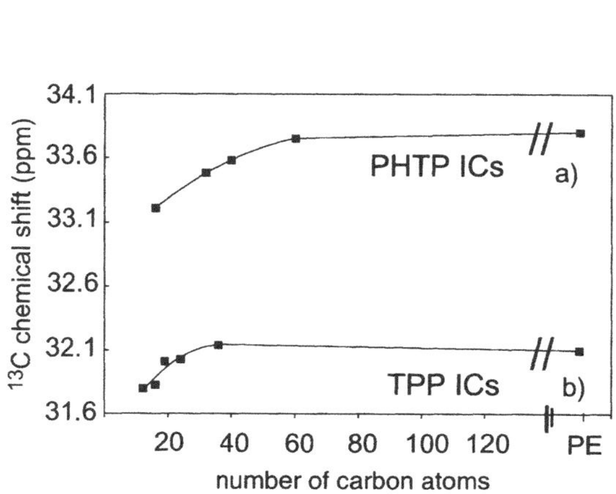

Fig. 17 Comparison of the inner methylene ^{13}C chemical shifts vs. number of carbons of *n*-alkanes and PE in (**a**) PHTP ICs and in (**b**) TPP ICs

5 Polyconjugated Oligomers Confined to Hexagonal Lattices

The strategy of imposing a constraining hexagonal lattice on the polymer chains has been applied to polyconjugated molecules for obtaining a special state of alignment of photoactive molecules in crystals [59–61]. In fact, linear π-conjugated molecules have been engineered in the unique supramolecular structures of TPP, in such a way as to be completely surrounded by aromatic groups. The molecules are confined by self-assembly within a crystalline honeycomb network and achieve uniform alignment along the crystallographic axis c, with stimulating perspectives of orienting anisotropically the active oligomers in thermally stable single-crystals. Advanced NMR techniques can localize with a rare accuracy the hydrogens involved in weak hydrogen–π interactions and found the central motif common to the novel nanostructures, that is the organization of π-receptors wrapping about polymer chains and sustaining a diffuse network of weak $\pi \cdots \pi$ and aliphatic CH$\cdots\pi$ interactions. Weak host–guest interactions form collectively a robust, exceptionally stable, architecture which nevertheless provides a soft environment for the guest: an ideal locus where the encapsulated molecules are balanced between freedom and constriction. The novel series of supramolecular adducts contribute to the evolution of new generations of 1D ordered assemblies of important families of functional oligomers with the strategy of exploiting weak interactions to modulate the molecular arrangements and functions.

2D NMR helped much deciding whether the macromolecules are massively included in the nanochannels and determining the specific interactions of the guests with the channel walls (Fig. 18).

The surprising upfield shifts due to magnetic susceptibility observed in this family of polymers embedded in aromatic nanochannels, and the 2D cross-peaks are clear demonstrations of the intimate relationship with the host aromatic groups. The occurrence of a strong effect of magnetic susceptibility, due to the TPP aromatic paddles, permits the exact determination of the distance of a probe atom on the chain from the phenylenedioxy ring (Fig. 15). Nucleus-independent chemical-shift maps give the calculated chemical shifts in the space region around the aromatic ring, as derived from the electronic current density generated by the main magnetic field [62]. Proton resonances shifted 2 ppm upfield indicate the topology of the guest hydrogens above the plane of benzene rings at a distance of 2.5 Å [63, 64]. Short intermolecular distances, such as those determined, imply close contact between the π-electron clouds and hydrogen atoms, and favorable Van der Waals interactions. At less than 3 Å distances of hydrogens from the center of aromatic rings, energy minima have been well established by both theoretical studies and a survey of a number of crystal structure determinations.

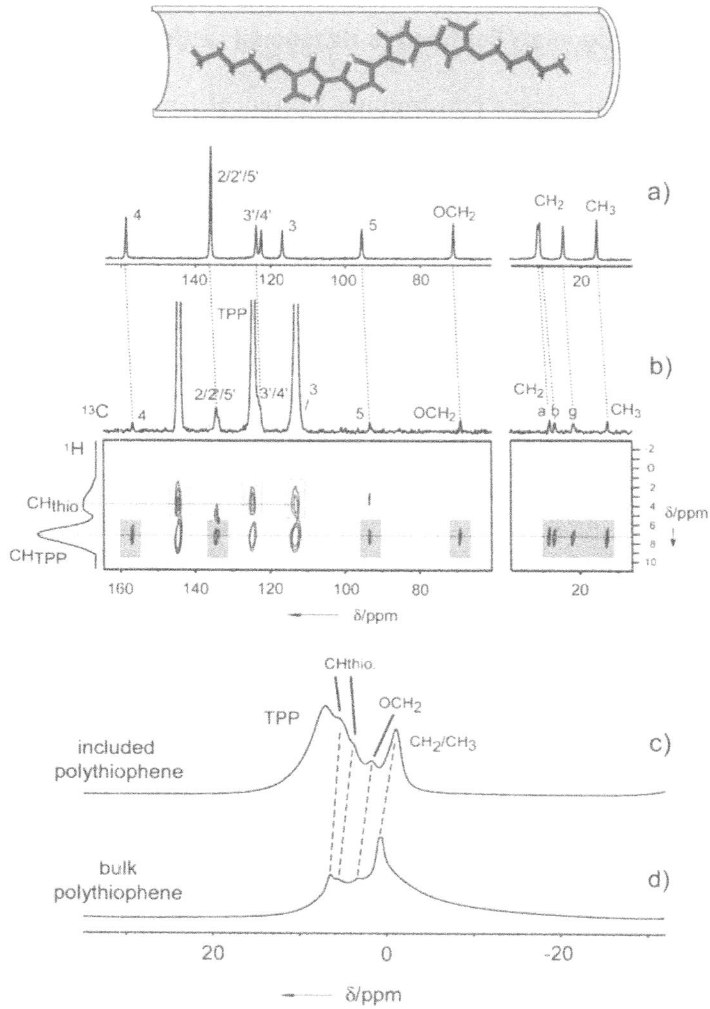

Fig. 18 (**a**) 1D ^{13}C Ramped-CP MAS NMR spectrum of the bulk 4,4‴-dipentoxy-2,2′:5′,2″:5″,2‴-tetrathiophene (**b**) Contour plot of 2D Phase-Modulated Lee–Goldburg heteronuclear (^{1}H–^{13}C) dipolar correlation spectrum of the polythiophene nanostructured material in the hexagonal lattice of TPP (**c**) and (**d**) ^{1}H MAS NMR spectra (15 kHz spinning speed) of included and bulk polythiophene show a large up-field shift of the resonance when the polymer is included in the aromatic nanochannel

The $\pi \cdots \pi$ arrangements account for about 2 kcal/mol of energy [65–67]. Recently, energy minima of 2.5 kcal/mol were accurately determined for benzene dimers in the favorable T-shaped and slipped-parallel arrangements. At the minima, the intermolecular distance of aromatic CH hydrogens from the center of the next aromatic ring is 2.5 Å, as in the present case [68]. The XRD structural resolution, although unable to detect the exact location of the polymer chains, describes the crystal host matrix where the guests are encapsulated. The existence of CH$\cdots\pi$ interactions is compatible with the size of the 5 Å nanochannels that are recognizable in the crystal structure. Three contiguous phenylenedioxy rings are present at each cross-section of the TPP structure (Fig. 15) and another three, related by a 6_3 screw axis, are found on the next TPP layer along the channel.

Oligothiophenes substituted by terminal alkoxy-chains exhibit a peculiar phenomena of differential mobility between the core and the chain ends. In fact, the chain ends present gauche conformations that slow down their motion. On the contrary, the thiophene core, appended to the alkoxy plugs, shows exceptionally high librational mobility in both the ground and excited states as a molecular gyroscope [60].

Thus, both the aromatic and aliphatic CH groups of the guest find with high probability any of the π-receptors lining the walls and move to neighboring receptors by low activation energies. Multiple CH$\cdots\pi$ interactions simultaneously sustain the architecture, and make a notable contribution to the exothermic transition of self-assembly. Interestingly, the infinite fully-aromatic nanochannels shaped by the host resembles carbon nanotubes with aromatic rings parallel to the channel axis.

6 Conclusions

The organization into the hexagonal lattice is closely related to a state of fast motion occurring on the main-chain or on articulated side chains of flexible macromolecules. Solid state NMR is particularly suited to the study of the peculiar arrangements both in bulk polymers and copolymers as well as in nanocomposite materials.

In particular, multinuclear solid-state NMR can describe in detail the role of chain-dynamics and weak interactions cooperating to fabricate polymer nanostructured materials organized in hexagonal host lattices that show exceptional thermal and chemical stability. The recognition of weak interactions in supramolecular assemblies of synthetic polymers can bring significant consequences to the study of polymers at interfaces and promote projects for obtaining innovative heterogeneous materials and nanocomposites. In addition, the structure and dynamics of the flexible polymer chain in the unusual hexagonal lattices can shed light on the intermolecular interactions in the bulk crystal state.

Acknowledgements The authors would like to thank the FISR program on nanocomposite materials and MIUR for financial support.

References

1. Brown SP, Spiess HW (2001) Chem Rev 101:4125
2. Bovey FA, Mirau PA (1996) NMR of Polymers. Academic, San Diego
3. Comotti A, Simonutti R, Bracco S, Castellani L, Sozzani P (2001) Macromolecules 34:4879
4. Alamo RG, VanderHart DL, Nyden MR, Mandelkern L (2000) Macromolecules 33:6094
5. Kuwabara K, Horii F (1999) Macromolecules 32:5600
6. Callaghan PT, Samulski ET (2003) Macromolecules 33:3795
7. Cheng HN, English AD (2003) ACS Symp Ser 834:3
8. Bassi IW, Corradini P, Fagherazzi G, Valvassori A (1970) Eur Polym J 6709
9. Shirayama K, Kita S, Watabe H (1972) Makrom Chem 151:97
10. De Ballesteros OR, Auriemma F, Guerra G, Corradini P (1996) Macromolecules 29:7141
11. Kraack H, Deutsch M, Sirota EB (2000) Macromolecules 33:6174
12. Keller A, Cheng SZD (1998) Polymer 39:4461
13. de Langen M, Prins KO (1999) Chem Phys Lett 299:195
14. de Langen M, Prins KO (2000) Polymer 41:1175
15. Hu WG, Boeffel C, Schmidt-Rohr K (1999) Macromolecules 32:1611
16. Yamakawa H, Matsukawa S, Kurosu H, Kuroki S (1999) J Chem Phys 111:7110
17. Stewart MJ, Jarrett WL, Mathias LJ, Alamo RG, Mandelkern L (1996) Macromolecules 29:4963
18. Möller M, Cantow H-J, Drotloff H, Emeis D, Lee K-S, Wegner G (1986) Makromol Chem 137:1237
19. Ishikawa S, Ando I (1991) J Mol Struct 273:227
20. Yamakawa H, Matsukawa S, Kurosu H, Kuroi S, Ando I (1998) Chem Phys Lett 283:333
21. Hu W, Sirota EB (2003) Macromolecules 36:5144
22. Bracco S, Comotti A, Simonutti R, Camurati I, Sozzani P (2002) Macromolecules 35:1677
23. Hu W, Srinivas S, Sirota EB (2002) Macromolecules 35:5013
24. Möller M (1988) Makromol Chem Rapid Commun 9:107
25. Sozzani P, Behling RW, Schilling FC, Bruckner S, Helfand E, Bovey FA, Jelinski LW (1989) Macromolecules 22:3318
26. Crosby RC, Haw JF (1987) Macromolecules 20:2324
27. Alexander MN, Desper CR, Sagalyn PL, Schneider NS (1977) Macromolecules 10:721
28. Meille SV, Farina A, Gallazzi MC, Sozzani P, Simonutti R, Comotti A (1995) Macromolecules 28:1893
29. Simonutti R, Veeman WS, Ruhnau FC, Gallazzi MC, Sozzani P (1996) Macromolecules 29:4958
30. Simonutti R, Comotti A, Sozzani P (1996) J Inorg Organomet Polym 6:313
31. Corradi E, Farina A, Gallazzi MC, Brückner S, Meille SV (1999) Polymer 40:4473
32. MacNicol DD, Toda F, Bishop R (1996) (eds) Solid-State Supramolecular Chemistry: Crystal Engineering, vol 6. Pergamon, UK
33. Farina M, Allegra G, Natta G (1964) J Am Chem Soc 86:2516
34. Hollingsworth MD, Harris KDM (1996) Urea, thiourea, and selenourea. Comprehensive Supramolecular Chemistry, vol. 6. Pergamon, UK, pp 177–237
35. Cannarozzi GM, Meresi GH, Vold RL, Vold RR (1991) J Phys Chem 95:1525
36. Sozzani P, Bovey FA, Schilling FC (1989) Macromolecules 22:4225
37. Schilling FC, Bovey FA, Sozzani P (1990) Polymer Preprints 31

38. Schilling FC, Bovey FA, Sozzani P (1991) Macromolecules 24:4369
39. Sozzani P, Bovey FA, Schilling FC (1991) Macromolecules 24:6764
40. Schilling FC, Amundson KR, Sozzani P (1994) Macromolecules 27:6498
41. Comotti A, Simonutti R, Catel G, Sozzani P (1999) Chem Mater 11:1476
42. Sozzani P, Comotti A, Bracco S, Simonutti R (2004) Chem Comm 767
43. Harada A, Jun L, Kamachi M (1994) Nature 370:126
44. Harada A (2001) Accounts Chem Res 34:456
45. Lu J, Mirau PA, Tonelli AE (2001) Macromolecules 34:3276
46. Simonutti R, Sozzani P, Bracco S, Comotti A (2001) Polym Mater Sci Eng 82:161
47. Farina M, Di Silvestro G, Grassi M (1979) Makromol Chem 180:1041
48. Farina M, Di Silvestro G, Sozzani P (1996) Perhydrotriphenylene: a D3 symmetric host. Comprehensive Supramolecular Chemistry, vol. 6. Pergamon, UK, pp 371–398
49. Brückner S, Sozzani P, Boeffel C, Destri S, Di Silvestro G (1989) Macromolecules 22:607
50. Tonelli AE (1990) Macromolecules 23:3134
51. Tonelli AE (1990) Macromolecules 23:3129
52. Sozzani P, Bracco S, Comotti A, Ferretti L, Simonutti R (2005) Angew Chem Int Ed 44:1816
53. Vinogradov SP, Madhu PK, Vega S (1999) Chem Phys Lett 314:443
54. Sozzani P, Bracco S, Comotti A, Simonutti R, Camurati I (2003) J Am Chem Soc 125:12881
55. Sozzani P, Simonutti R, Bracco S, Comotti A (2003) Polymer Preprints 44:297
56. Haliloglu T, Mattice WL (1993) Macromolecules 26:3137
57. Badjic JD, Balzani V, Credi A, Silvi S, Stoddart JF (2004) Science 303:1845
58. De Gennes PG, Prost J (1993) The Physics of Liquid Crystals, 2nd edn. Science Publication.Clarendon, Oxford
59. Sozzani P, Comotti A, Bracco S, Simonutti R (2004) Angew Chem Int Ed 43:2792
60. Brustolon M, Barbon A, Bortolus M, Maniero AL, Sozzani P, Comotti A, Simonutti R (2004) J Am Chem Soc 126:15512
61. Barbon A, Bortolus M, Brustolon M, Comotti A, Maniero AL, Segre U, Sozzani P (2003) J Phys Chem B 107:3325
62. Von Ragué Schleyer P, Maerker C, Dransfeld A, Jiao H, Van Eikema Hommes NJR (1996) J Am Chem Soc 118:6317
63. Waugh JS, Fessenden RW (1957) J Am Chem Soc 79:846
64. Rapp A, Schnell I, Sebastiani D, Brown SP, Percec V, Spiess HW (2003) J Am Chem Soc 125:13284
65. Jeffrey GA (1997) An Introduction to Hydrogen Bonding. Oxford University Press, New York
66. Nishio M (2004) Cryst Eng Comm 6:130
67. Desiraju GR (2002) Acc Chem Res 35:565
68. Tsuzuki S, Honda K, Uchimaru T, Mikami M, Tanabe K (2002) J Am Chem Soc 124:104

Author Index Volumes 101–181

Author Index Volumes 1–100 see Volume 100

Subject Index

The manufacturer's authorised representative in the EU is Springer Nature Customer Service Centre GmbH, Europaplatz 3, 69115 Heidelberg, Germany. If you have any concerns regarding our products, please contact ProductSafety@springernature.com

Printed and bound by CPI Group (UK) Ltd, Croydon, CR0 4YY
07/01/2026
02031056-0005